The Fence Methods

The Fence Methods

Jiming Jiang
University of California, Davis, USA

Thuan Nguyen
Oregon Health & Science University, USA

World Scientific

NEW JERSEY · LONDON · SINGAPORE · BEIJING · SHANGHAI · HONG KONG · TAIPEI · CHENNAI · TOKYO

Published by

World Scientific Publishing Co. Pte. Ltd.

5 Toh Tuck Link, Singapore 596224

USA office: 27 Warren Street, Suite 401-402, Hackensack, NJ 07601

UK office: 57 Shelton Street, Covent Garden, London WC2H 9HE

Library of Congress Cataloging-in-Publication Data
Jiang, Jiming.
   The fence methods / by Jiming Jiang (University of California, Davis, USA), Thuan Nguyen (Oregon Health & Science University, USA).
     pages cm
   Includes bibliographical references and index.
   ISBN 978-9814596060 (hardcover : alk. paper)
   1. Partially ordered sets. 2. Group theory. 3. Multiple criteria decision making. I. Nguyen, Thuan (Professor of biostatistics) II. Title.
   QA171.485.J53 2015
   519.5--dc23

                                              2015026036

British Library Cataloguing-in-Publication Data
A catalogue record for this book is available from the British Library.

To beloved children living in poverty

Preface

The past two decades have seen an explosion of interest in statistical model selection that is largely driven by practical needs. The traditional methods of model selection, a core of which are the information criteria, encounter difficulties in dealing with high-dimensional and complex problems.

The main difficulty brought by the high-dimensional problems is computational. For example, in regression variable selection when the number of candidate variable, p, is large, it is computationally expensive, or even infeasible, to carry out all-subset selections, as required by the information criteria. Furthermore, when p is larger than n, the sample size, the standard method of fitting the least squares, which is needed in computing the information criterion function, is not possible. There have been major breakthroughs in high-dimensional model selection, thanks to the proposal of shrinkage selection/estimation methods [Tibshirani (1996), Fan and Li (2001), among others].

On the other hand, complex selection problems are often encountered. For example, in many problems of practical interest, the observations are correlated in complex ways. Mixed effects models, such as linear and generalized linear mixed models, have been used in analyzing such complex data, but little was known about model selection in such situations. In particular, the use of information criteria was largely *ad hoc* for mixed model selection. In some other cases, the distribution of data is not fully specified. Here, once again, one encounters difficulty in using the information criteria, because the likelihood function is not available. Another complication came when the data involve missing values, which occurs frequently in practice. The standard methods for model selection, including associated software package, were built for the complete-data situations. As such, these methods and software cannot be directly applied to cases of missing or incomplete data.

In a breakthrough in complex model selection, Jiang, Rao, Gu and Nguyen [Jiang *et al.* (2008)] proposed a new class of strategies for model selection, which they coined *fence methods*. The idea consists of a procedure to isolate a subgroup of candidate models, known as "correct models", by building a statistical fence, or barrier. Once the fence is built, an optimal model is selected from those within the fence according to a criterion that can be made flexible. In particular, the criterion of optimality can take practical considerations into account. A number of variations of the fence have since been developed. The fence was motivated by the need to overcome the difficulties encountered by the information criteria. Major features of the fence include (i) flexibility in choosing both the measure of lack-of-fit and the criterion of optimality for selection within the fence; (ii) it is data-driven, giving the data plenty of opportunities to "speak out" in making some difficult decisions, namely, the choice of tuning parameters; and (iii) it leaves a room for practical considerations that is specialized to the current problem. A recent review by Jiang (2014) has provided an overview of the fence, including major advances, applications, and open problems. A software package, recently developed by T. Nguyen, J. Zhao, J. S. Rao and J. Jiang and available online at http://fencemethods.com/, has implemented most of the methods associated with the fence.

This monograph is devoted to giving a detailed account of the fence, including its variations, and related topics. It is mainly intended for researchers and graduate students, at M.S. or higher level. The monograph is mostly self-contained. A first course in mathematical statistics, the ability to use computer for data analysis, and familiarity with calculus and linear algebra are prerequisites.

Our research on complex model selection was first initiated by Dr. J. Sunil Rao, who brought up the problem of mixed model selection in the late 1990s. The collaborative research between Dr. Rao and Dr. Jiming Jiang has led to the 2008 paper that, for the first time, introduced the fence methods. Part of the topics presented in the monograph is based on the Ph.D. dissertation by Dr. Thuan Nguyen, who has made important contributions to the development of the fence. Other former students who have contributed, at various points, to the developments include former Ph.D. students Zhonghua Gu, Jiani Mu, Senke Chen, and Erin Melcon, and former M.S. students Mei-Chin Lin, Jianyang Zhao, Xi Ai, Xiaoyun Wang, and Haomiao Meng. In addition, we would like to thank Qui Tran for initiating the LaTex typesetting for the monograph, and Pete Scully and Michael Lin for helping with the design of the front cover. Our thanks also go to Professors Alan Welsh and Samuel Müller, who invited the authors to visit their institutes in Australia in 2013 that has led to many constructive discussions, especially related to the fence methods. We also thank Professors Welsh and Müller, Professors Partha Lahiri, Danny Pfeffermann, and J. N. K. Rao, and Dr. Long Ngo for their comments, and encouragement, on the fence methods, either in their published articles and books or through personal communications.

Jiming Jiang and Thuan Nguyen
Davis, California and Portland, Oregon
March 2015

Contents

Chapter 1

Introduction

On the morning of March 16, 1971, Hirotugu Akaike, as he was taking a seat on a commuter train, came out with the idea of a connection between the relative Kullback-Liebler discrepancy and the empirical log-likelihood function, a procedure that was later named Akaike's information criterion, or AIC [Akaike (1973, 1974); see Bozdogan (1994) for the historical note]. The idea has allowed major advances in model selection and related fields. See, for example, de Leeuw (1992).

To introduce the idea of a new model selection strategy, it is important to understand the "old" strategies, at the center of which are the information criteria. Below we provide a brief review of the criteria. But before we do this, let us keep in mind one of the best-known quotes in Statistics, or perhaps all of Science. George Box, one of the most influential statisticians of the 20th century, once wrote that "essentially, all models are wrong, but some are useful." What it means is that, even though there may not exist a "true" model, in reality, a suitable choice of one may still provide a good (or, perhaps, the best) approximation, from a practical standpoint.

1.1 The information criteria

Suppose that one wishes to approximate an unknown probability density function (pdf), g, by a given pdf, f. The Kullback–Leibler (K-L) discrepancy, or information, defined as

$$I(g; f) = \int g(x) \log g(x) \, dx - \int g(x) \log f(x) \, dx, \qquad (1.1)$$

provides a measure of lack of approximation. It can be shown, by Jensen's inequality, that the K-L information is always nonnegative, and it equals zero if and only if $f = g$ a.e. [i.e., $f(x) = g(x)$ for all x except on a set of Lebesgue measure zero]. However, K-L information is not a distance (Exercise 1.1). Note that the first term on the right side of (1.1) does not depend on f. Therefore, to best approximate g, one needs to find an f that minimizes $-\int g(x) \log f(x) \, dx = -E_g\{\log f(X)\}$, where E_g means that the expectation is taken with $X \sim g$. Since we do not know g, the expectation is not computable. However, suppose that we have independent observations $X_1, \ldots, X_n$ from g. Then we may replace the expectation by the sample mean, $n^{-1} \sum_{i=1}^{n} \log f(X_i)$, which is an unbiased estimator for the expectation. In particular, under a parametric model, denoted by M, the pdf f depends on a vector θ_M of parameters, denoted by $f = f_M(\cdot|\theta_M)$. For example, in a linear regression model, M may correspond to a subset of predictors, and θ_M the vector of corresponding regression coefficients. Then the AIC is a two-step procedure. The first step is to find the θ_M that minimizes

$$-\frac{1}{n} \sum_{i=1}^{n} \log f_M(X_i|\theta_M) \tag{1.2}$$

for any given M. Note that (1.2) is simply the negative log-likelihood function under M. Therefore, the θ_M that minimizes (1.2) is the maximum likelihood estimator (MLE), denoted by $\hat{\theta}_M$. Then, the second step of AIC is to find the model M that minimizes

$$-\frac{1}{n} \sum_{i=1}^{n} \log f_M(X_i|\hat{\theta}_M). \tag{1.3}$$

However, there is a serious drawback in this approach: Expression (1.3) is no longer an unbiased estimator for $-E_g\{\log f_M(X|\theta_M)\}$ due to overfitting. The latter is caused by double-use of the same data—for estimating the expected log-likelihood and for estimating the parameter vector θ_M. Akaike (1973) proposed to rectify this problem by correcting the bias, which is

$$\frac{1}{n} \sum_{i=1}^{n} E_g\{\log f_M(X_i|\hat{\theta}_M)\} - E_g\{\log f_M(X|\theta_M)\}.$$

He showed that, asymptotically, the bias can be approximated by $|M|/n$, where $|M|$ denotes the dimension of M defined as the number of estimated parameters under M. For example, if M is an ARMA(p,q) model in time series [e.g., Shumway (1988)], then $|M| = p + q + 1$ (the 1 corresponds to the unknown variance). Thus, a term $|M|/n$ is added to (1.3), leading to

$$-\frac{1}{n} \sum_{i=1}^{n} \log f_M(X_i|\hat{\theta}_M) + \frac{|M|}{n}.$$

The expression is then multiplied by the factor $2n$, which does not depend and affect the choice of M, to come up with the AIC:

$$\text{AIC}(M) = -2 \sum_{i=1}^{n} \log f_M(X_i | \hat{\theta}_M) + 2|M|. \tag{1.4}$$

In words, the AIC is minus two times the maximized log-likelihood plus two times the number of estimated parameters.

A number of similar criteria have been proposed since the AIC. These include the Bayesian information criterion [BIC; Schwarz (1978)], and a criterion due to Hannan and Quinn (1979). All of these criteria may be expressed, in a general form, as

$$\text{GIC}(M) = \hat{D}_M + \lambda_n |M|, \tag{1.5}$$

where $\hat{D}_M$ is a measure of lack-of-fit by the model M and λ_n is a penalty for complexity of the model, which may depend on the effective sample size, n. The measure of lack-of-fit is such that a model of greater complexity fits better, therefore has a smaller $\hat{D}_M$; on the other hand, such a model receives more penalty for having a larger $|M|$. The effective sample size is equal to the sample size, if the samples are i.i.d.; otherwise, it might not be the same as the sample size (see below). Therefore, criterion (1.5), known as the generalized information criterion, or GIC [Nishii (1984); Shibata (1984)], is a trade-off between model fit and model complexity. In particular, the AIC corresponds to (1.5) with $\hat{D}_M$ being -2 times the maximized log-likelihood and $\lambda_n = 2$; the BIC and HQ have the same $\hat{D}_M$, but $\lambda_n = \log n$ and $c \log \log n$, respectively, where c is a constant greater than 2. We consider another special cases below.

Example 1.1: Hurvich and Tsai (1989) argued that in the case of the ARMA(p, q) model, a better bias correction could be obtained if one replaces $p + q + 1$ by an asymptotically equivalent quantity, $n(p+q+1)/(n - p - q - 2)$. This leads to a modified criterion known as AICC. The AICC corresponds to (1.5) with $\lambda_n = 2n/(n - p - q - 2)$. So, if $n \to \infty$ while the ranges of p and q are bounded, AICC is asymptotically equivalent to AIC.

One concern about AIC is that it does not lead to consistent model selection if the dimension of the optimal model is finite. Here, an optimal model is defined as a correct model with minimum dimension. For example, suppose that the true underlying model is AR(2); then AR(3) is a correct model (by letting the coefficient corresponding to tha additional term equal to zero), but not an optimal model. On the other hand, AR(1) is an incorrect model, or wrong model [that the true underlying model is AR(2) implies that the true coefficient corresponding to the second-order term is nonzero]. So, if one considers all AR models as candidates, the only optimal model is AR(2). Furthermore, consistency of model selection is defined as that the probability of selecting an optimal model goes to 1 as $n \to \infty$.

On the other hand, the BIC and HG are consistent model selection procedures. One may wonder what causes such a difference. The idea is quite simple, and it has something to do with the choice of λ_n. The AIC is not consistent because it has not given enough penalty for complex models. For example, suppose that the true underlying model is AR(p). Then AIC tends to choose an order higher than p in selecting the order for the AR model. This problem is called overfitting. It can be shown that AIC does not have the other kind of problem, namely, underfitting, meaning that the procedure tends to select an order less than p. In other words, asymptotically, AIC is expected to select, at least, a correct model; but the selected model may not be optimal in that it can be further simplified. For the same reason, AICC is not consistent.

For a procedure to be consistent, one needs to control both overfitting and underfitting. Thus, on the one hand, one needs to increase the penalty λ_n in order to reduce overfitting; on the other hand, one cannot overdo this; otherwise, the underfitting will again make the procedure inconsistent. The question then is: What is the "right" amount of penalty? As far as the consistency is concerned, this is determined by the order of λ_n. It turns out that BIC and HG have the right order for λ_n that guarantees the consistency. See, for example, Jiang (2010) for further explanation.

Furthermore, the lack of consistency does not necessarily imply that AIC is inferior to BIC or HQ, from a practical standpoint. The reason is that the concept of consistency, as defined above, applies only to the "ideal" situation, where the true underlying model is of finite dimension, and is among the candidate models. In practice, however, such an ideal situation almost never occurs (remember "all models are wrong"). What one has instead is a collection of candidate models as approximation to the true underlying model, which is not one of the candidates. For example, in time series analysis, one may use an AR(p) model as an approximation to the true underlying model, which may be expressed as AR(∞). In such a case, it may be argued that the BIC and HQ are inconsistent, while the AIC is consistent in the sense that the selected order (of the AR model) tends to infinity as the sample size increases, which approximates the order of the true model.

1.2 Difficulties with the information criteria

Although the information criteria are broadly used, difficulties are often encountered, especially in some non-conventional situations. We discuss a number of such cases below.

1. The effective sample size. As mentioned, the λ_n that is involved in the information criteria, (1.5), may depend on n, which is supposed to be the effective sample size. For example, if the data are i.i.d., the effective sample size should be the same as the sample size, because every new observation provides, in a way, the same amount of new information. On the other hand, if all the data points are identical, the effective sample size should be 1, regardless of the number of observations, because every new observation provides no additional information. Of course, the latter case is a bit extreme, but there are many practical situations where the observations are correlated, even though they are not identical. One of those situations is mixed effects models. We illustrate with a simple example.

Example 1.2: Consider a linear mixed model defined as $y_{ij} = x'_{ij}\beta + u_i + v_j + e_{ij}$, $i = 1,\ldots,m_1$, $j = 1,\ldots,m_2$, where x_{ij} is a vector of known covariates, β is a vector of unknown regression coefficients (the fixed effects), u_i, v_j are random effects, and e_{ij} is an additional error. It is assumed that u_i's, v_j's and e_{ij}'s are independent, and that, for the moment, $u_i \sim N(0,\sigma_u^2)$, $v_j \sim N(0,\sigma_v^2)$, $e_{ij} \sim N(0,\sigma_e^2)$. It is well-known (e.g., Harville 1977, Miller 1977) that, in this case, the effective sample size for estimating σ_u^2 and σ_v^2 is not the total sample size $m_1 \cdot m_2$, but m_1 and m_2, respectively, for σ_u^2 and σ_v^2. Now suppose that one wishes to select the fixed covariates, which are components of x_{ij}, under the assumed model structure, using the BIC. It is not clear what should be in place of $\lambda_n = \log n$. For example, it does not make sense to let $n = m_1 \cdot m_2$.

2. The dimension of a model. Not only the effective sample size, the dimension of a model, $|M|$, can also cause difficulties. In some cases, such as the ordinary linear regression, this is simply the number of parameters under M, but in other situations where nonlinear, adaptive models are fitted, this can be substantially different. Ye (1998) developed the concept of generalized degrees of freedom (gdf) to track model complexity. A computational algorithm at heart, the method simply repeats the model fitting on perturbed values of the response, y (via resampling), and observes how the fitted values, $\hat{y}$, change. The sum of the sensitivities to change across all of the observations provides an approximation to the model complexity. It can be shown that, in the case of ordinary linear regression, this results in the number of parameters in the model. On the other hand, in the case of multivariate adaptive regression splines [MARS; Friedman (1991)], k nonlinear terms can have an effect of approximately $3k$ degrees of freedom. As another example, for classification and regression trees [CART; Breiman *et al.* (1984)], regression trees in ten dimensional noise each split costs approximately 15 degrees of freedom. While a general algorithm in its essence, the gdf approach requires significant computations. It is not at all clear how a plug-in of gdf for $|M|$ in (1.5) will affect the selection performance of the criteria.

3. Unknown distribution. In many cases, the distribution of the data is not fully specified (up to a number of unknown parameters); as a result, the likelihood function is not available. For example, suppose that normality is not assumed in Example 1.2 for the random effects and errors. In fact, the only distributional assumptions are that the random effects and errors are independent, have means zero and variances σ_u^2, σ_v^2 and σ_e^2, respectively. Such a model is referred to as a non-Gaussian linear mixed model [Jiang (2007)]. Again, suppose that one wishes to select the fixed covariates using AIC, BIC, or HQ. It is not clear how to do this because the likelihood is unknown under the assumed model. Of course, one could still blindly use those criteria, pretending that the data are normal, but the criteria are no longer what they mean to be. For example, Akaike's bias approximation that led to the AIC, as discussed in Section 1.1, is no longer valid.

4. Finite-sample performance, and the effect of a constant. Even in the conventional situation, there are still practical issues regarding the use of these criteria. For example, the BIC is known to have the tendency of overly penalizing "bigger" models. In other words, the penalizer, $\lambda_n = \log n$, may be a little too much in some cases. In such a case, one may wish to replace the penalizer by $c \log(n)$, where c is a constant less than one. Question is: What c? Asymptotically, the choice of c does not make a difference in terms of consistency of model selection, so long as $c > 0$. However, practically, it does. For example, comparing BIC with HQ, the penalizer of the latter is lighter in its order ($\log n$ vs $\log \log n$), but there is a constant c involved in HQ, which is supposed to be greater than 2. However, if $n = 100$, we have $\log n \approx 4.6$ and $\log \log n \approx 1.5$, hence, if the constant c in HQ is chosen as 3, BIC and HQ are almost the same.

In fact, there have been a number of modifications of the BIC aiming at improving the finite-sample performance. For example, Broman and Speed (2002) proposed a δ-BIC method by replacing the λ_n in BIC, which is $\log n$, by $\delta \log n$, where δ is a constant carefully chosen to optimize the finite-sample performance. However, the choice of δ relies on (extensive) Monte-Carlo simulations, and is case-by-case. For example, the value of δ depends on the sample size. Therefore, it is not easy to generalize the δ-BIC method. As another example, in our earlier experience we had studied the performance of the GIC (see Section 1.1) in selecting mixed logistic models, which are special cases of the generalized linear mixed models [GLMMs; e.g., Jiang (2007)]. We found that the finite-sample performance of the GIC is very sensitive to the choice of a "tuning constant". In other words, if the λ_n in (1.5) is replaced by $c\lambda_n$, where c is the tuning constant, then, depending on the choice of c, the performace of the GIC is very different, especially when the sample size is relatively small (Exercise 1.2).

    5. Criterion of optimality. Strictly speaking, model selection is hardly a purely statistical problem–it is usually associated with a problem of practical interest. Therefore, it seems a bit unnatural to let the criterion of optimality in model selection be determined by purely statistical considerations, such as the likelihood and K-L information. Other considerations, such as scientific and economic concerns, need to be taken into account. For example, what if the optimal model selected by the AIC is not to the best interest of a practitioner, say, an economist? In the latter case, can the economist change one of the selected variables, and do so "legitimately"? Furthermore, the dimension of a model, $|M|$, is used to balance the model complexity through (1.5). However, the minimum-dimension criterion, also known as *parsimony*, is not always as important. For example, the criterion of optimality may be quite different if prediction is of main interest.

    It is these concerns, such as the above, that led to the development of a new class of strategies for model selection, to be introduced in the sequel.

1.3   The fence method

As noted, in model selection problems one often deals with a trade-off between model fit and model complexity. It is a trade-off because the two usually head in opposite directions, that is, the improvement of model fit is usually at the cost of increasing model complexity. However, there are other situations, in statistics, where such dilemmas have to be dealt with, and have been dealt with. One of these is hypothesis testing. In the latter case, the type I and type II errors are known to be going in opposite directions; more importantly, there is a strategy to deal with it, through the control of the type I error. Namely, a level of significance, α, is chosen to control the (probability of) type I error. The goal is then to find a test that minimizes the (probability of) type II error, that is, most powerful, among all of the tests whose type I errors are under the control.

One may consider using a similar strategy to deal with the trade-off between model fit and complexity. First we need a measure for the model fit. This is known as a measure of lack-of-fit, defined as $Q(M, \theta_M; y)$, where M corresponds to a candidate model, θ_M a vector of parameters under M, and y the data vector. If the measure is minimized over the parameter space, Θ_M, one gets $Q(M; y) = \inf_{\theta_M \in \Theta_M} Q(M, \theta_M; y)$. The Q is a measure of lack-of-fit if (i) its value is nonnegative; (ii) if M_1 is a submodel of M_2 (i.e., M_1 is a special case of M_2), then $Q(M_1; y) \geq Q(M_2; y)$ for any y; and (iii) $\mathrm{E}\{Q(M, \theta_M; y)\}$ is minimized when M is a true model, and θ_M the true parameter vector under M. Here by true model we mean that M is a correct model in the sense that, for some θ_M, the pair (M, θ_M) identifies the true underlying distribution of y. Throughout this monograph, we use the terms "true model" and "correct model" interchangeably. However, being a true model not necessarily imply that it is the most efficient one. For example, in regression model selection, a true model is one that contains at least all the variables whose coefficients are non-zero, but the model remains true if an additional variable is added, whose coefficient is zero. Below are some examples of measures of lack-of-fit.

Example 1.3: (Negative log-likelihood) Suppose that the joint distribution of y belongs to a family of parametric distributions $\{P_{M,\theta_M}, M \in \mathcal{M}, \theta_M \in \Theta_M\}$. Suppose that P_{M,θ_M} have a (joint) probability density function (pdf), $f_M(\cdot|\theta_M)$ with respect to a σ-finite measure, μ. Consider

$$Q(M, \theta_M; y) = -\log\{f_M(y|\theta_M)\}, \tag{1.6}$$

the negative log-likelihood function. It is easy to show that this is defines a measure of lack-of-fit (Exercise 1.3).

Example 1.4: (AR model selection) A special class of time series models is the autoregressive model, or AR model, defined by $y_t - b_1 y_{t-1} - \cdots - b_p y_{t-p} = \epsilon_t$, assumed to hold for all $t = 0, \pm 1, \pm 2, \ldots$, where ϵ_t is a white noise process that are independent with mean zero and (unknown) constant variance. If the polynomial $b(z) = 1 - b_1 z - \cdots - b_p z^p$ satisfies $b(z) \neq 0$ for all complex numbers z such that $|z| \leq 1$, then $\{y_t\}$ is stationary time series. Note that, in this case, a model M is determined by its order, p, such that the corresponding coefficients, $b_1, \ldots, b_p$ satisfies $b(z) \neq 0, |z| \leq 1$. Suppose that one observes $y_1, \ldots, y_T$, where $T > P$, where P is the maximum order under consideration. Define

$Q(M, \theta_M; y) = \sum_{t=P+1}^{T} (y_t - b_1 y_{t-1} - \cdots - b_p y_{t-p})^2$. It can be shown that this defines a measure of lack-of-fit (Exercise 1.4).

Classical model selection assumes the existence of a true underlying model that is among the set of candidate models, denoted by $\mathcal{M}$. Although this assumption is necessary in establishing consistency of the model selection, it limits the scope of applications. In practice, a true model simply may not exist, or exist but not among $\mathcal{M}$ (once again, recall that "all models are wrong"). Furthermore, in many cases, the definition of a "model" is, by far, not as clear as in the traditional sense. In particular, the distribution of y may not be fully specified (up to a vector of unknown parameters), under the assumed model. For such reasons, we need to extend the above definitions of true model and true parameter vector, as well as measure of lack-of-fit.

Under the extended definition, we do not assume the existence of a true model in $\mathcal{M}$. Instead, a vector $\theta_M^* \in \Theta_M$ is called a true parameter vector under M with respect to Q if it minimizes $\mathrm{E}\{Q(M, \theta_M; y)\}$, that is,

$$\mathrm{E}\{Q(M, \theta_M^*; y)\} = \inf_{\theta_M \in \Theta_M} \mathrm{E}\{Q(M, \theta_M; y)\} \equiv Q(M), \qquad (1.7)$$

where the expectation is taken with respect to the true distribution of y (which does not depend on M but may be unknown). A true model by Q is a model $M \in \mathcal{M}$ such that

$$Q(M) = \inf_{M' \in \mathcal{M}} Q(M'). \qquad (1.8)$$

Note that here the true model by Q is defined as a model that provides the best approximation, or best fit to the data, which is not necessarily a true model in the traditional sense. However, the latter definitions are extensions of the traditional concepts defined earlier. The main difference is that, now a measure of lack-of-fit, Q, must satisfy (iii) under the extended definition of true model and true parameter vector. In other words, we can drop (iii) as a requirement for the measure of lack-of-fit, because it is automatically satisfied. We consider some more examples.

Example 1.3 (continued): A model M is called a true model, in the traditional sense, if $y \sim P_{M, \theta_M}$, that is, the (joint) distribution of y is P_{M, θ_M}, for some $\theta_M \in \Theta_M$. Such a θ_M is called a true parameter vector. It turns out, by the properties of the log-likelihood function, that if M is a true model, then $\theta_M^* = \theta_M$, the true parameter vector under M. Furthermore, by the same argument, it can be shown that $Q(M)$ is minimized when M is a true model. In other words, $\mathrm{E}\{Q(M, \theta_M; y)\}$ is minimized when M is a true model and θ_M a true parameter vector under M, a property required by (iii). Therefore, any true model in the traditional sense is a true model by Q for the Q defined by (1.6) (Exercise 1.3).

Example 1.5: (Residual sum of squares) Consider the problem of selecting the covariates for a linear model so that $E(y) = X\beta$, where X is a matrix of covariates whose columns are to be selected from a number of candidates $X_1, \ldots, X_K$ and β is a vector of regression coefficients. A candidate model M corresponds to $X_M \beta_M$, where the columns of X_M are a subset of $X_1, \ldots, X_K$ and β_M is a vector of regression coefficients of suitable dimension. Consider

$$Q(M, \beta_M; y) = |y - X_M \beta_M|^2, \tag{1.9}$$

known as the residual sum of squares (RSS). A model M is a true model if $E(y) = X_M \beta_M$ for some β_M, which is called a true vector of regression coefficients. It can be shown that the Q defined by (1.9) has the property that $E\{Q(M, \beta_M; y)\}$ is minimized when M is a true model and β_M a corresponding true vector of regression coefficients. It follows that if M is a true model that corresponds to X_M, then $\beta_M^* = \beta_M$, the true vector of regression coefficients corresponding to X_M; and any true model is a true model by Q for the Q defined by (1.9) (Exercise 1.5).

Example 1.6: (MVC model selection) If a model is only specified by the mean and covariance matrix of y, it is called a mean and variance/covariance model, or MVC model. In this case, one may consider

$$Q(M, \theta_M; y) = |(T'V_M^{-1}T)^{-1}T'V_M^{-1}(y - \mu_M)|^2, \tag{1.10}$$

where μ_M and V_M are the mean vector and covariance matrix under M, and T is a given $n \times s$ matrix of full rank $s \leq n$, where n is the dimension of y and s is a fixed integer. Here θ_M represents the vector of any parameters that are involved in μ_M and V_M; in particular, if μ_M, V_M are completely specified under M, θ_M can be dropped. It can be shown that the right side of (1.10) is minimized when $\mu_M = E(y)$, and $V_M = \mathrm{Cov}(y)$, where E and Cov represent the true expectation and covariance matrix, respectively. Thus, we have yet another measure of lack-of-fit defined by (1.10), under the extended definition (Exercise 1.6).

Suppose that a measure of lack-of-fit, Q, has been chosen. Let $\tilde{M}$ be a "baseline model" in the sense that $\tilde{M} \in \mathcal{M}$ and $Q(\tilde{M}) = \inf_{M \in \mathcal{M}} Q(M)$; in other words, $\tilde{M}$ is the optimal model in terms of the fit. Note that, in many cases, $\tilde{M}$ can be determined without any calculation. For example, if $\mathcal{M}$ contains a "full model", say M_f, that is, a model such that all the other models in $\mathcal{M}$ are submodels of M_f, then we have $\tilde{M} = M_f$ by the definition of the measure of lack-of-fit. Furthermore, if $\mathcal{M}$ contains a true model, then M_f is (also) a true model; however, it may not be the most efficient true model. To be more precise, suppose that, here, efficiency is defined in terms of dimesionality. Namely, if $Q(M_1) = Q(M_2)$ but $|M_1| < |M_2|$, M_1 is more efficient than M_2; in other words, if two models are the same in terms of the fit, but one model is of smaller dimension than the other, the model with small dimension is more efficient. Sometimes, one would rather tolerate a little in terms of the fit in exchange for a lower dimension. For example, suppose that, in regression variable selection, there are 100 candidate variables to be selected as predictors, so that $|M_f| = 100$. Furthermore, suppose that M_1, which consists of 3 of the candidate variables with $|M_1| = 3$, is almost as good as M_f in terms of fit, that is, $Q(M_1) \approx Q(M_f)$, then, clearly, the choice is M_1 rather than M_f. Here, a key point is that, once we know that two models are similar in terms of the fit, the choice is easy to make, according to the complexity, or dimensionality in this case—one does not need to compute a criterion function, like (1.5), in order to make a decision. This key observation leads to the following strategy: (I) set an upper bound for the difference in terms of the fit to isolate a subclass of models, known as the "correct" models; (II) within the subclass of correct models, choose the one that has the smallest dimension. Mathematically, (I) can be written as

$$Q(M) - Q(\tilde{M}) \leq b, \tag{1.11}$$

where b represents a bound for the distance, in terms of Q, between a candidate model, M, and the baseline model, $\tilde{M}$, that is considered "good enough", or sufficient. Note that the left side of (1.11) is always nonnegative. Intuitively, the subclass of models isolated by (1.11) are those that are within "striking distance" from the baseline model in terms of the fit, and from this subclass we choose the model that is the simplest.

The question then is: What b? Note that the role of b is meant to separate the correct models from the incorrect ones. Therefore, to answer the question, it would help if we have some idea about that the left side of (1.11) is likely to be when M is a correct model, and how it might be different when M is incorrect. A heuristic argument shows that, if M, $\tilde{M}$ are both correct, the difference $Q(M) - Q(\tilde{M})$ is in the order of its standard deviation (s.d.); otherwise, the difference is expected to be of (much) higher order. We illustrate this with an example.

Example 1.3 (continued): Suppose that $\mathcal{M}$ contains a true model as well as a full model, M_f, so that $\tilde{M} = M_f$. It turns out that M_f is a true model. If M is another true model, but $|M| < |M_f|$, the difference $2\{Q(M) - Q(\tilde{M})\}$ is the likelihood-ratio (LR) statistic, which has an asymptotic $\chi^2_{|M_f| - |M|}$ distribution (e.g., Lehmann 1999, p. 527). Thus, the left side of (1.11) is $O_P(1)$, the same order as the asymptotic s.d., $\sqrt{(|M_f| - |M|)/2}$, if $|M_f|$ is bounded. On the other hand, if M is an incorrect model, the order of $Q(M) - Q(\tilde{M})$ is much higher. To see this, consider a special case in which the observations, $y_1, \ldots, y_n$, are i.i.d, and $y = (y_i)_{1 \le i \le n}$. Then, for any fixed θ_M, we have

$$Q(M, \theta_M; y) - Q(\tilde{M}) \ge Q(M, \theta_M; y) - Q(M_f, \theta_f; y) = \sum_{i=1}^{n} Y_i, \qquad (1.12)$$

where θ_f is the true parameter vector under M_f, and

$$Y_i = \log\{f_{M_f}(y_i|\theta_f)/f_M(y_i|\theta_M)\}.$$

Note that $Y_1, \ldots, Y_n$ are i.i.d. with $E(Y_i) > 0$ (Exercise 1.7). Therefore, the right side of (1.12) is $O_P(n)$, and this is true for every fixed θ_M. Thus, by taking the infimum over $\theta_M \in \Theta_M$, one expects the left side of (1.11) also to be $O_P(n)$ (although the conclusion does requires some regularity conditions).

Let $s(M, \tilde{M}) = \text{s.e.}\{Q(M) - Q(\tilde{M})\}$ be the standard error of $Q(M) - Q(\tilde{M})$, that is, an estimate of the s.d. of $Q(M) - Q(\tilde{M})$. Then, a reasonable choice for the b in (1.11) is $b = s(M, \tilde{M})$. The quantity serves as a "fence" to confine the correct models and exclude the incorrect ones. For such a reason, the procedure is called *fence*. Note that b depends on M. It should also be noted that the construction of the fence only completes the first step of the fence, that is, step (I), but the second step, that is, the selection of a model within the fence, is often straightforward. Nevertheless, the second step is one of the features that distinguishes the fence from other model selection procedures. Namely, the criterion used to select the model within the fence is flexible, and can incorporate practical considerations. In particular, the minimum dimension criterion is not necessarily the one used to selection the model within the fence. An economist may prefer one model over the other, even though the preferred model is of higher dimension–the economist is free to make the choice, because the model is in the fence; then it is up to the economist to make the "rules" on which model to choose.

When the minimum-dimension rule is used in selecting the model within the fence, the fence procedure can be simplified as follows. Let $d_1 < d_2 < \cdots < d_L$ be all the different dimensions of the models in $\mathcal{M}$, here assumed to be a finite collection.

The fence algorithm:

i) Find $\tilde{M}$.

ii) Compute $s(M, \tilde{M})$ for all $M \in \mathcal{M}$ such that $|M| = d_1$; let $\mathcal{M}_1 = \{M \in \mathcal{M} : |M| = d_1$ and (1.11) holds with $b = s(M, \tilde{M})\}$; if $\mathcal{M}_1 \neq \emptyset$, stop (no need for any more computation!). Let M_0 be the model in $\mathcal{M}_1$ such that $Q(M_0) = \min_{M \in \mathcal{M}_1} Q(M)$; M_0 is the selected model.

iii) If $\mathcal{M}_1 = \emptyset$, compute $s(M, \tilde{M})$ for all $M \in \mathcal{M}$ such that $|M| = d_2$; let $\mathcal{M}_2 = \{M \in \mathcal{M} : |M| = d_2$ and (1.11) holds with $b = s(M, \tilde{M})\}$; if $\mathcal{M}_2 \neq \emptyset$, stop. Let M_0 be the model in $\mathcal{M}_2$ such that $Q(M_0) = \min_{M \in \mathcal{M}_2} Q(M)$; M_0 is the selected model.

iv) Continue until the procedure stops (Exercise 1.8).

In short, the algorithm may be described as follows: Check the candidate models, from the simplest to the most complex. Once one has discovered a model that falls within the fence and checked all the other models of the same simplicity (for their membership within the fence), one stops.

1.4 Evaluation of $s(M, \tilde{M})$

As mentioned, in the case of negative log-likelihood measure of lack-of-fit (e.g., Example 1.3), which from now on shall be called ML model selection, with $\tilde{M} = M_{\mathrm{f}}$, one may use $s(M, \tilde{M}) = \sqrt{(|M_{\mathrm{f}}| - |M|)/2}$ as the cut-off b in (1.11). However, if measures other than the negative log-likelihood is used, this cut-off may not be appropriate.

In the following, we consider a few other special cases. It should be pointed out that consistent estimators are not necessarily available under a candidate model, M, unless M is a true model. Nevertheless, the asymptotic behavior of the fence is mainly determined by the order of $s(M, \tilde{M})$, so it is not really necessary to estimate the variance, or s.d., of $Q(M) - Q(\tilde{M})$ consistently, although it would be better if such an estimator is available.

1.4.1 *Clustered observations*

Clustered observations arise naturally in many fields, including analysis of longitudinal data (e.g., Diggle *et al.* (2002) and small area estimation (e.g., Rao (2003)). Let $y_i = (y_{ij})_{1 \leq j \leq k_i}$ represent the vector of observations in the ith cluster, and $y = (y_i)_{1 \leq i \leq m}$. We assume that $y_1, \ldots, y_m$ are independent. Furthermore, we assume that Q is *additive* in the sense that

$$Q(M, \theta_M; y) = \sum_{i=1}^{m} Q_i(M, \theta_M; y_i). \qquad (1.13)$$

We consider some examples.

Example 1.3 (continued): Under the independence assumption, we have $f_M(y|\theta_M) = \prod_{i=1}^{m} f_{M,i}(y_i|\theta_M)$, where $f_{M,i}(\cdot|\theta_M)$ is the joint pdf of y_i under M and θ_M, we have $Q(M, \theta_M; y) = -\sum_{i=1}^{m} \log\{f_{M,i}(y_i|\theta_M)\}$. Thus, (1.13) holds with $Q_i(M, \theta_M; y_i) = -\log\{f_{M,i}(y_i|\theta_M)\}$.

Example 1.6 (continued): Consider the MVC model selection. Let $T = \mathrm{diag}(T_1, \ldots T_m)$, where T_i is $k_i \times s_i$ and $1 \leq s_i \leq k_i$, we have $Q(M, \theta_M; y) = \sum_{i=1}^{m} |(T_i'V_{M,i}^{-1}T_i)^{-1}T_i'V_{M,i}^{-1}(y_i - \mu_{M,i})|^2$, where $\mu_{M,i}$ and $V_{M,i}$ are the mean vector and covariance matrix of y_i under M and θ_M, respectively. Thus, (1.13) holds with $Q_i(M, \theta_M; y_i) = |(T_i'V_{M,i}^{-1}T_i)^{-1}T_i'V_{M,i}^{-1}(y_i - \mu_{M,i})|^2$.

Example 1.7: (Extended GLMM selection) Jiang and Zhang (2001) proposed an extension of the generalized linear mixed models [GLMM; e.g., Jiang (2007)], in which only the conditional mean of the response given the random effects is parametrically specified. It is assumed that, given a vector α of random effects, the responses $y_1, \ldots, y_n$ are conditionally independent such that $\mathrm{E}(y_i|\alpha) = h(x_i'\beta + z_i'\alpha)$, $1 \leq i \leq n$, where $h(\cdot)$ is a known function, β is a vector of unknown fixed effects, and x_i, z_i are known vectors. Furthermore, it is assumed that $\alpha \sim N(0, \Sigma)$, where the covariance matrix Σ depends on a vector ψ of variance components. Let β_M and ψ_M denote β and ψ under M, and $g_{M,i}(\beta_M, \psi_M) = \mathrm{E}\{h_M(x_i'\beta_M + z_i'\Sigma_M^{1/2}\xi)\}$, where h_M is the function h under M, Σ_M is the covariance matrix under M evaluated at ψ_M, and the expectation is taken with respect to $\xi \sim N(0, I_m)$ (which does not depend on M). Here m is the dimension of α and I_m the m-dimensional identity matrix. We consider

$$Q(M, \theta_M; y) = \sum_{i=1}^{n} \{y_i - g_{M,i}(\beta_M, \psi_M)\}^2. \qquad (1.14)$$

It is easy to see that the Q given above satisfies the basic requirement, that is, conditions (i)–(iii) given in the second paragraph of Section 1.3, as a measure of lack-of-fit (Exercise 1.9). In fact, (1.14) corresponds to the Q in Example 1.6 with $T = I$, the identity matrix. Note that, since V is not parametrically specified under the assumed model, it should not get involved in the Q. Therefore, (1.14) is a natural choice in this case. Furthermore, the measure (1.14) is additive (even if the data is not clustered).

Suppose that the measure Q is additive. It is easy to show that

$$\text{var}\{Q(M, \theta_M; y) - Q(\tilde{M}, \theta_{\tilde{M}}; y)\}$$

$$= \text{E}\left\{\sum_{i=1}^{m}(Q_i - \tilde{Q}_i)^2 - \sum_{i=1}^{m}(\mu_i - \tilde{\mu}_i)^2\right\}, \tag{1.15}$$

where $Q_i = Q_i(M, \theta_M; y_i)$, $\tilde{Q}_i = Q_i(\tilde{M}, \theta_{\tilde{M}}; y_i)$, $\mu_i = \text{E}(Q_i)$, and $\tilde{\mu}_i = \text{E}(\tilde{Q}_i)$. Thus, $s^2(M, \tilde{M})$ can be approximated by the so-called *observed variance*, that is, by removing the expectation sign on the right side of (1.15), replacing θ_M by $\tilde{\theta}_M$, the minimizer of $Q(M, \theta_M; y)$ over $\theta_M \in \Theta_M$, and replacing μ_i by its estimator, if available; similarly for $\theta_{\tilde{M}}$ and $\tilde{\mu}_i$.

1.4.2 *Gaussian models*

A Gaussian model, M, is characterized by its mean vector, μ_M, and covariance matrix, V_M, hence Gaussian model selection is all about selecting μ_M and V_M. For example, a Gaussian mixed model can be expressed as $y = X\beta + Z\alpha + \epsilon$, where X is a matrix of known covariates, β is a vector of unknown fixed effects, Z is a known matrix, α is a vector of random effects, and ϵ is a vector of errors. It is assumed that α and ϵ are jointly normally distributed with $\text{Var}(\alpha) = G$, $\text{Var}(\epsilon) = R$ and $\text{cov}(\alpha, \epsilon) = 0$, where G and R are the covariance matrices under the assumed model. It is clear that Gaussian mixed model is a special case of Gaussian model with $\mu_M = X_M \beta_M$ and $V_M = R_M + Z_M G_M Z'_M$, where X_M, β_M, Z_M, G_M and R_M are the corresponding matrices or vector under model M. It can be shown that, for fixed $\theta_M, \theta_{\tilde{M}}$, we have

$$\text{var}\{Q(M, \theta_M; y) - Q(\tilde{M}, \theta_{\tilde{M}}; y)\}$$

$$= \frac{1}{2}\text{tr}\{(V_M^{-1}V_{\tilde{M}} - I)^2\}$$

$$+ (\mu_M - \mu_{\tilde{M}})'V_M^{-1}V_{\tilde{M}}V_M^{-1}(\mu_M - \mu_{\tilde{M}}), \tag{1.16}$$

if Q is the negative log-likelihood (Example 1.3); and

$$\text{var}\{Q(M, \theta_M; y) - Q(\tilde{M}, \theta_{\tilde{M}}; y)\}$$

$$= 2\left(\text{tr}\left[\{(T'V_M^{-1}T)^{-2}T'V_M^{-1}V_{\tilde{M}}V_M^{-1}T\}^2\right] - \text{tr}\left\{(T'V_{\tilde{M}}^{-1}T)^{-2}\right\}\right)$$

$$+ 4(\mu_M - \mu_{\tilde{M}})'C_M V_{\tilde{M}} C_M(\mu_M - \mu_{\tilde{M}}), \tag{1.17}$$

where $C_M = V_M^{-1}T(T'V_M^{-1}T)^{-2}T'V_M^{-1}$, if Q is given by (1.10) for the MVC model selection (Exercise 1.10). Thus, one may obtain $s^2(M, \tilde{M})$ by replacing μ_M, V_M, $\mu_{\tilde{M}}$ and $V_{\tilde{M}}$ on the right sides of (1.16), (1.17) by $\tilde{\mu}_M$, $\tilde{V}_M$, $\tilde{\mu}_{\tilde{M}}$ and $\tilde{V}_{\tilde{M}}$, respectively, where $\tilde{\mu}_M$ is μ_M with θ_M replaced by $\tilde{\theta}_M$, etc., where $\tilde{\theta}_M$ is the minimizer of $Q(M, \theta_M; y)$ over $\theta_M \in \Theta_M$.

1.4.3 *Non-Gaussian linear mixed models*

Consider a non-Gaussian linear mixed model [e.g., Jiang (2007)]. Let y denote the vector of observations. Suppose that

$$y = X\beta + Z_1\alpha_1 + \cdots + Z_s\alpha_s + \epsilon, \qquad (1.18)$$

where X is a known matrix of covariates; β is an unknown vector of fixed effects; $Z_1, \ldots, Z_s$ are known matrices; $\alpha_1, \ldots, \alpha_s$ are vectors of random effects, and ϵ is a vector of errors. It is assumed that, for each $1 \leq r \leq s$, the components of α_r are independent and distributed as $N(0, \sigma_r^2)$; the components of ϵ are independent and distributed as $N(0, \tau^2)$, and $\alpha_1, \ldots, \alpha_s, \epsilon$ are independent. Note that normality is not assumed for the random effects and errors, neither is any other specific distribution. On the other hand, the left side of (1.17) typically involve higher (3rd and 4th) moments of the random effects and errors, which are not functions of $\sigma_r^2, 1 \leq r \leq s$ and τ^2, if the data are not normal. However, we may use a method developed by Jiang (2005).

First introduce some notation. Let $f(i_1, \ldots, i_4) = \sum_{r=0}^{s} \kappa_r z_{i_1 r} \cdots z_{i_4 r}$, where $\kappa_r = \mathrm{E}(\alpha_{r1}^4) - 3\sigma_r^4$ is the kurtosis of α_{r1}, the first component of α_r, $1 \leq r \leq s$, and κ_0 is the kurtosis of ϵ_1, the first component of ϵ, defined similarly; $z_{ir}' = (z_{irl})_{1 \leq l \leq m_r}$ is the ith row of Z_r, with $Z_0 = I_n$, $m_r = \dim(\alpha_r), 1 \leq r \leq s$ and $m_0 = n = \dim(\epsilon)$; and $z_{i_1 r} \cdots z_{i_4 r} = \sum_{l=1}^{m_r} z_{i_1 rl} \cdots z_{i_4 rl}$. Define $\Gamma(i_1, i_2) = \sum_{r=0}^{s} \gamma_r z_{i_1 r} \cdot z_{i_2 r}$, where $\gamma_r = \sigma_r^2/\tau^2, 1 \leq r \leq s$, and $\gamma_0 = 1$, and $z_{i_1 r} \cdot z_{i_2 r} = \sum_{l=1}^{m_r} z_{i_1 rl} z_{i_2 rl}$. Let $f_1, \ldots, f_L$ be the different functional values of $f(i_1, \ldots, i_4)$ (i.e., different functions). Write $u = y - X\beta = (u_i)_{1 \leq i \leq n}$. Let $A = (a_{ij})$ and $B = (b_{ij})$ be symmetric matrices. Define

$$c(i_1, \ldots, i_4) = \frac{\sum_{f(i_1, \ldots, i_4) = f_l} a_{i_1 i_2} b_{i_3 i_4}}{|\{f(i_1, \ldots, i_4) = f_l\}|}, \quad \text{if } f(i_1, \ldots, i_4) = f_l,$$

$1 \leq l \leq L$, where $|\cdot|$ denotes the cardinality. Similarly, let

$$t(i_1, i_2, i_3) = \mathrm{E}(u_{i_1} u_{i_2} u_{i_3}) = \sum_{r=0}^{s} \mathrm{E}(\alpha_{r1}^3) z_{i_1 r} \cdot z_{i_2 r} \cdot z_{i_3 r},$$

where $z_{i_1 r} \cdot z_{i_2 r} \cdot z_{i_3 r} = \sum_{l=1}^{m_r} z_{i_1 rl} z_{i_2 rl} z_{i_3 rl}$. Let $t_1, \ldots, t_K$ be the different nonzero functional values of $t(i_1, i_2, i_3)$. Let $b = (b_i)$ be a vector. Define

$$c(i_1, i_2, i_3) = \frac{\sum_{t(i_1, i_2, i_3) = t_l} a_{i_1 i_2} b_{i_3}}{|\{t(i_1, i_2, i_3) = t_l\}|}, \quad \text{if } t(i_1, i_2, i_3) = t_l,$$

$1 \leq l \leq K$. The following results are established in Jiang (2005).

Lemma 1.1. *Under the non-Gaussian linear mixed model, we have*

$$\text{cov}(u'Au, u'Bu)$$

$$= \text{E}\left\{ \sum_{f(i_1,\ldots,i_4)\neq 0} c(i_1,\ldots,i_4)u_{i_1}\cdots u_{i_4} \right\}$$

$$+ \left\{ 2\text{tr}(AVBV) - 3\tau^4 \sum_{f(i_1,\ldots,i_4)\neq 0} c(i_1,\ldots,i_4)\Gamma(i_1,i_3)\Gamma(i_2,i_4) \right\},$$

$$\text{cov}(u'Au, b'u) = \text{E}\left\{ \sum_{i_1,i_2,i_3} c(i_1,i_2,i_3)u_{i_1}u_{i_2}u_{i_3} \right\} \qquad (1.19)$$

for any symmetric matrices A, B, where $V = \text{Var}(y)$.

Consider, for example, the MVC model selection of Example 1.6. Let $\tilde{M} = M_f$ and assume that $\mathcal{M}$ contains a true model. It is easy to show that $Q(M, \theta_M; y) = (y - \mu)'C_M(y - \mu) + 2(\mu - \mu_M)'C_M(y - \mu) + c_M$, where C_M is defined below (1.17), and c_M is nonrandom. Thus, we have

$$Q(M) - Q(\tilde{M}) = (y - \mu)'(C_M - C_{\tilde{M}})(y - \mu)$$
$$+ 2\{(\mu - \mu_M)'C_M - (\mu - \mu_{\tilde{M}})'C_{\tilde{M}}\}(y - \mu) + c_M - c_{\tilde{M}}$$
$$= I_1 + 2I_2 + c_M - c_{\tilde{M}}.$$

It follows that the left side of (1.17) is equal to $\text{var}(I_1) + 4\text{cov}(I_1, I_2) + 4\text{var}(I_2)$. By Lemma 1.1, we have $\text{var}(I_1) = v_{11} + v_{12}$,

$$v_{11} = \text{E}\left\{ \sum_{f(i_1,\ldots,i_4)\neq 0} c(i_1,\ldots,i_4)u_{i_1}\cdots u_{i_4} \right\},$$

and

$$v_{12} = 2\text{tr}[\{(C_M - C_{\tilde{M}})V\}^2] - 3\tau^4 \sum_{f(i_1,\ldots,i_4)\neq 0} c(i_1,\ldots,i_4)\Gamma(i_1,i_3)\Gamma(i_2,i_4),$$

with $c(i_1,\ldots,i_4)$ given as in Lemma 1.1 with $A = B = C_M - C_{\tilde{M}}$ and $f(i_1,\ldots,i_4)$ determined under $\tilde{M}$. Note that the latter is a true model.

Similar to Jiang (2005), an estimator of $\text{var}(I_1)$ is obtained as $\widetilde{\text{var}}(I_1) = \tilde{v}_{11} + \tilde{v}_{12}$, where $\tilde{v}_{1j}$, $j = 1, 2$ are defined as follows: (i) $\tilde{v}_{11}$ is v_{11} with the expectation sign removed, C_M evaluated under M and $\tilde{\theta}_M$, and everything else under $\tilde{M}$ and $\tilde{\theta}_{\tilde{M}}$. Using the terms of Jiang (2005), the estimator $\widetilde{\text{var}}(I_1)$ is called a *partially observed* variance, with $\tilde{v}_{11}$ being the observed part and $\tilde{v}_{12}$ the estimated part.

Similarly, an expression of $\text{cov}(I_1, I_2)$ is obtained by Lemma 1.1 with $A = C_M - C_{\tilde{M}}$, $b = C_M(\mu_{\tilde{M}} - \mu_M)$ and $t(i_1, i_2, i_3)$ determined under $\tilde{M}$. Unlike the previous case, the estimator of $\text{cov}(I_1, I_2)$, $\widetilde{\text{cov}}(I_1, I_2)$, consists entirely of an observed covariance, given by the right side of (1.19) with the expectation sign removed, C_M and μ_M evaluated under M and $\hat{\theta}_M$, and everything else under $\tilde{M}$ and $\hat{\theta}_{\tilde{M}}$.

Finally, since $\text{var}(I_2) = (\mu_M - \mu_{\tilde{M}})'C_M V C_M(\mu_M - \mu_{\tilde{M}})$, it can be estimated by $\widetilde{\text{var}}(I_2) = (\tilde{\mu}_M - \tilde{\mu}_{\tilde{M}})'\tilde{C}_M \tilde{V}_{\tilde{M}} \tilde{C}_M(\tilde{\mu}_M - \tilde{\mu}_{\tilde{M}})$.

In conclusion, $s^2(M, \tilde{M})$ is given by $\widetilde{\text{var}}(I_1) + 4\widetilde{\text{cov}}(I_1, I_2) + 4\widetilde{\text{var}}(I_2)$.

1.4.4   *Extended GLMMs*

Consider the extended GLMM of Example 1.7. To obtain an expression for $s(M, \tilde{M})$, write $\xi_{M,i} = g_{M,i}^2(\beta_M, \psi_M) - 2y_i g_{M,i}(\beta_M, \psi_M)$, $\xi_{\tilde{M},i} = \xi_{M,i}$ with M replaced by $\tilde{M}$, $d_i = \xi_{M,i} - \xi_{\tilde{M},i}$, and $\delta_i = g_{M,i}(\beta_M, \psi_M) - g_{\tilde{M},i}(\beta_{\tilde{M}}, \psi_{\tilde{M}})$. Here θ_M is the vector that minimizes $\text{E}(Q_M)$ over Θ_M, the parameter space for θ_M, and $\theta_{\tilde{M}}$ the true vector of parameters under $\tilde{M}$, assuming that $\tilde{M}$ is a true model. It can be shown that, under regularity conditions, the left side of (1.17) is equal to $\text{var}(\sum_{i=1}^n d_i)\{1 + o(1)\}$ and

$$
\text{var}\left(\sum_{i=1}^n d_i\right)
$$

$$
= 4\left\{\text{E}\left(\sum_{i=1}^n \delta_i^2 y_i^2\right) + \sum_{i \neq j} \delta_i \delta_j g_{\tilde{M},i,j}(\beta_{\tilde{M}}, \psi_{\tilde{M}})1_{(z_i' \Sigma_{\tilde{M}} z_j \neq 0)}\right.
$$

$$
\left. - \sum_{i,j} \delta_i \delta_j g_{\tilde{M},i}(\beta_{\tilde{M}}, \psi_{\tilde{M}}) g_{\tilde{M},j}(\beta_{\tilde{M}}, \psi_{\tilde{M}})1_{(z_i' \Sigma_{\tilde{M}} z_j \neq 0)}\right\},
$$

where $g_{\tilde{M},i,j}(\beta_{\tilde{M}}, \psi_{\tilde{M}}) = \text{E}\{h_{\tilde{M}}(x_i'\beta_{\tilde{M}} + z_i'\Sigma_{\tilde{M}}^{1/2}\xi)h_{\tilde{M}}(x_j'\beta_{\tilde{M}} + z_j'\Sigma_{\tilde{M}}^{1/2}\xi)\}$ with $\xi \sim N(0, I_m)$. Thus, $s^2(M, \tilde{M})$ is obtained as a *partially observed variance*,

$$
4\left\{\sum_{i=1}^n \tilde{\delta}_i^2 y_i^2 + \sum_{i \neq j} \tilde{\delta}_i \tilde{\delta}_j g_{\tilde{M},i,j}(\tilde{\beta}_{\tilde{M}}, \tilde{\psi}_{\tilde{M}})1_{(z_i'\tilde{\Sigma}_{\tilde{M}} z_j \neq 0)}\right.
$$

$$
\left. - \sum_{i \neq j} \tilde{\delta}_i \tilde{\delta}_j g_{\tilde{M},i}(\tilde{\beta}_{\tilde{M}}, \tilde{\psi}_{\tilde{M}}) g_{\tilde{M},j}(\tilde{\beta}_{\tilde{M}}, \tilde{\psi}_{\tilde{M}})1_{(z_i'\tilde{\Sigma}_{\tilde{M}} z_j \neq 0)}\right\},
$$

where $\tilde{\delta}_i$ is δ_i with $\beta_M, \psi_M, \beta_{\tilde{M}}$ and $\psi_{\tilde{M}}$ replaced by $\tilde{\beta}_M, \tilde{\psi}_M, \tilde{\beta}_{\tilde{M}}$ and $\tilde{\psi}_{\tilde{M}}$, respectively, and $\tilde{\Sigma}_{\tilde{M}}$ is $\Sigma_{\tilde{M}}$ with $\psi_{\tilde{M}}$ replaced by $\tilde{\psi}_{\tilde{M}}$.

Remark. It should be noted that a good evaluation of $s(M, \tilde{M})$ is, in general, still difficult to obtain, especially when the observations are highly correlated. Even if a reasonable assessment is available, the computation of $s(M, \tilde{M})$ is usually much more time-consuming than that of $Q(M) - Q(\tilde{M})$. This could be a major issue for high dimensional selection problems. In Chapter 3, we take a different approach and show that, in a way, the evaluation of $s(M, \tilde{M})$ can be avoided.

1.5 A stepwise fence procedure

As indicated by the fence algorithm (end of Section 1.3), the procedure has the computational advantage that it starts with the simplest models and therefore may not need to search the entire model space in order to determine the optimal model. On the other hand, such a procedure may still involve a lot of evaluations when the model space is large. For example, in quantitative trait loci mapping, variance components arising from the trait genes, polygenic and environmental effects are often used to model the covariance structure of the phenotypes given the identity by descent sharing matrix (e.g., Almasy and Blangero (1998)). Such a model is usually complex due to the large number of putative trait loci. To make the fence procedure computationally more attractive to large and complex models, we propose the following variation of fence for situations of complex models with many predictors. Another variation of the fence aiming at (greatly) improving the computational efficiency is introduced in Chapter 5.

Broman and Speed (2002) proposed three stepwise procedures to avoid all-subset selection in their proposed δ-BIC procedure (see §1.2.4, namely, the forward selection (FW), the backward elimination (BW), and the forward-backward (F-B) procedure. The idea is to use one of the above sequential selection method to first generate a sequence of candidate models. The δ-BIC is then applied to the sequence, which is a subset of $\mathcal{M}$ in order to select the optimal model. In their simulation studies, the authors showed that the F-B procedure performs the best compared to FW and BW. Thus, we focus on the F-B and develop a similar fence procedure.

To be more specific, consider the extended GLMM of Example 1.7. Write $X = (x_i')_{1 \leq i \leq n}$ and $Z = (z_i')_{1 \leq i \leq n}$. We assume that there is a collection of covariate vectors $X_1, \ldots, X_K$, from which the columns of X are to be selected. Furthermore, we assume that there is a collection of matrices $Z_1, \ldots, Z_L$ such that $Z\alpha = \sum_{s \in S} Z_s \alpha_s$, where $S \subset \{1, \ldots, L\}$, and each

α_s is a vector of i.i.d. random effects with mean 0 and variance σ_s^2. The subset S is subject to selection. The parameters under an extended GLMM are the fixed effects and variances of the random effects. Note that in this case the full model corresponding to $X\beta + Z\alpha = \sum_{k=1}^{K} X_k \beta_k + \sum_{l=1}^{L} Z_l \alpha_l$ is among the candidate models. Thus, we let $\tilde{M} = M_f$. As noted, the idea is to use a F-B procedure to generate a sequence of candidate models, from which the optimal model is selected using the fence method. We begin with a FW selection. Let M_1 be the model that minimizes $Q(M)$ among all models with a single parameter; if M_1 is within the fence, stop the FW; otherwise, let M_2 be the model that minimizes $Q(M)$ among all models that add one more parameter to M_1; if M_2 is within the fence, stop the FW; and so on. The FW stops when the first model is discovered within the fence (eventually it will, because the full model is in the fence). The procedure is then followed by a BW elimination. Let M_k be the final model of the FW. If no submodel of M_k with one less parameter is within the fence, M_k is selected; otherwise, M_k is replaced by M_{k+1}, which is a submodel of M_k with one less parameter and is within the fence, and so on. We call such a variation of fence the F-B fence.

1.6   Summary and remarks

In this chapter, we introduced the fence method that is motivated by a number of difficulties associated with the traditional information criteria. We outlined a numerical procedure known as the fence algorithm. We discussed the evaluation of a standard error for the difference between measures of lack-of-fit, which is used as the cut-off in the fence inequality. A F-B fence procedure is proposed to reduce the computational burden for high-dimensional selection problems.

It should be noted that, at this point, the cut-off for the fence, that is, $b = s(M, \tilde{M})$ in (1.11), is by no means optimal. For example, one may consider, instead of (1.11),

$$Q(M) - Q(\tilde{M}) \leq c \cdot s(M, \tilde{M}), \tag{1.20}$$

where c is a constant different than 1. Here is another way to look at it. Consider $d = \{Q(M) - Q(\tilde{M})\}/s(M, \tilde{M})$, which may be regarded as the standardized difference. Then, what (1.11) suggests is that $d \leq c$ with $c = 1$; in other words, one is using, essentially, 1 s.d. as the cut-off. The question is: Is one s.d. really enough? Or should one consider

2 s.d., or 1.1 s.d? Some examples of the latter case are discussed in the next chapter. Note that, here, we are mainly concerned about the finite-sample performance, because, asymptotically, the choice of c does not make a difference in terms of the consistency of the model selection. The optimal choice of c, known as the *tuning constant*, will be discussed in Chapter 3.

Talking about the consistency, there have been a series of studies on the consistency of the fence, F-B fence, and other variations of the fence to be introduced in the sequel, as well as other related topics. We shall consider these asymptotic properties near the end of this monograph.

1.7 Exercises

1.1. Show that the K-L information defined by (1.1) is ≥ 0 with equality holding if and only if $f = g$ a.e.; that is, $f(x) = g(x)$ for all $x \notin A$, where A has Lebesgue measure zero. However, it is not a distance. Here a distance, $\rho(\cdot, \cdot)$, must satisfy the follwing for all x, y, z: (i) $\rho(x, y) \geq 0$; (ii) $\rho(x, y) = \rho(y, x)$; (iii) $\rho(x, z) \leq \rho(x, y) + \rho(y, z)$; and (iv) $\rho(x, y) = 0$ if and only if $x = y$.

1.2. This exercise may be regarded as a small research project. Consider a mixed logistic model such that, given the random effects $\alpha_1, \ldots, \alpha_m$, responses $y_{ij}, i = 1, \ldots, m, j = 1, \ldots, k$ are independent such that, with $\alpha = (\alpha_i)_{1 \leq i \leq m}$, we have $y_{ij}|\alpha \sim \text{Bernoulli}(p_{ij})$ and $\text{logit}(p_{ij}) = x'_{ij}\beta + \alpha_i$, where $\text{logit}(p) = \log\{p/(1-p)\}, p \in (0, 1)$. Furthermore, the random effects are independent and distributed as $N(0, \sigma^2)$.

Here, the first column of $X = (x'_{ij})_{1 \leq i \leq m}$ is 1_m, the $m \times 1$ vector of ones, corresponding to an intercept; the rest of the columns of X are to be X_1, X_2, X_3, X_4, where the components of X_1 and X_2 are generated independently from the Uniform$[0, 1]$ distribution; the components of X_3 and X_4 are generated independently from the Bernoulli(0.5) and Bernoulli(0.2) distributions, respectively. Let the true model be the one with $\beta = (1, 0.5, 0, -0.5, 0)'$ and $\sigma^2 = 0.25$ (or $\sigma = 0.5$). Thus, in particular, the true model involves X_1 and X_3 only.

a. Develop a numerical algorithm to evaluate the log-likelihood function under the mixed logistic model (Hint: First, you need to express the log-likelihood as a sum of terms that involve some one-dimensional integrals; then find a way to approximate those integrals).

b. Carry out a Monte-Carlo simulation investigating the performance of the δ-BIC (see §1.2.3) for different choice of δ. You may consider the sample sizes $m = 20, 40, 60$ and $k = 5$.

1.3. Show that the negative log-likelihood measure defined in Example 1.3 is a measure of lack-of-fit according to the definition above Example 1.3.

1.4. Show that the RSS measure defined in Example 1.4 is a measure of lack-of-fit according to the definition above Example 1.3.

1.5. Consider the measure Q defined by (1.9). Show that $E\{Q(M, \beta_M; y)\}$ is minimized when M is a true model and β_M a corresponding true vector of regression coefficients according to the definition in Example 1.5. It follows that if M is a true model that corresponds to X_M, then $\beta_M^* = \beta_M$, the true vector of regression coefficients corresponding to X_M; and any true model is a true model by Q for the Q defined by (1.9).

1.6. Consider the MVC model selection of Example 1.6. Show that the right side of (1.10) is minimized when $\mu_M = E(y)$, and $V_M = Cov(y)$, where E and Cov represent the true expectation and covariance matrix, respectively.

1.7. Show that the Y_i defined below (1.12) satisfies $E(Y_i) = E(Y_1) > 0$.

1.8. Show that, in the fence algorithm (given at the end of Section 1.3), the procedure must stop at some point.

1.9. Show that, in Example 1.7, (1.14) is a measure of lack-of-fit, that is, it satisfies conditions (i)–(iii) given in the second paragraph of Section 1.3. In particular, show that $E\{Q(M, \theta_M; y)\}$ is minimized when M is a true model and $\theta_M = (\beta_M', \psi_M')'$ is the true parameter vector under M.

1.10. Derive the expressions (1.16) and (1.17).

Chapter 2

Examples

Before we move on to introduce some of the most important advances in the fence methods, we would like to use a number of examples, including simulations and real data analyses, to demonstrate the fence procedures as well as the scope of potential applications.

2.1 Examples with simulations

This section consists of five simulated examples.

2.1.1 *Regression model selection*

Under a nonlinear regression model, the observations $y_1, \ldots, y_n$ satisfy

$$y_i = g_M(x_{M,i}, \beta_M) + \epsilon_i,$$

where M indicates the assumed model, $g_M(\cdot, \cdot)$ is a known function, $x_{M,i}$ is a vector of known covariates, β_M is a vector of unknown regression coefficients, and ϵ_i is an error. It is assumed that $\epsilon_1, \ldots, \epsilon_n$ are independent with mean 0 and variance σ^2. The model selection problem is thus equivalent to selecting the mean function $\mu_M = (\mu_{M,i})_{1 \leq i \leq n}$, where $\mu_{M,i} = g_M(x_{M,i}, \beta_M)$.

A special case is variable selection in linear regression. In this case, there are a collection of candidate covariate vectors, say, $X_1, \ldots, X_K$, where $X_j = (x_{ij})_{1 \leq i \leq n}$, such that $g_M(x_{M,i}, \beta_M) = \sum_{j \in S} \beta_{M,j} x_{ij}$, where S is a subset of $\{1, \ldots, K\}$ determined by M, and $\beta_M = (\beta_{M,j})_{j \in S}$.

For regression model selection, an example of $Q(M)$ is given by

$$Q(M) = \sum_{i=1}^{n} \{y_i - g_M(x_{M,i}, \beta_M)\}^2. \tag{2.1}$$

To see that Q is a measure of lack-of-fit, let M^* be a true model and β_{M^*} the corresponding true vector of regression coefficients. Then, we have

$$y_i - g_M(x_{M,i}, \beta_M)$$
$$= y_i - g_{M^*}(x_{M^*,i}, \beta_{M^*}) + g_{M^*}(x_{M^*,i}, \beta_{M^*}) - g_M(x_{M,i}, \beta_M)$$
$$= \epsilon_i + d_i(M, M^*).$$

It follows that $E\{Q(M)\} = n\sigma^2 + \sum_{i=1}^n d_i^2(M, M^*) \geq n\sigma^2$. On the other hand, if M is a true model and β_M the corresponding true vector of regression coefficients, we must have $E\{Q(M)\} = n\sigma^2$.

Next, we examine the performance of the fence method through a simulation study using a similar design as Shao and Rao (2000). We consider a set of candidate covariate vectors $X_1, \ldots, X_K$, where $X_j = (x_{ij})_{1 \leq i \leq n}$, such that $g_M(x_{M,i}, \beta_M) = \sum_{j \in S} \beta_{M,j} x_{ij}$, where S is a subset of $\{1, \ldots, K\}$ determined by M, and $\beta_M = (\beta_{M,j})_{j \in S}$. Here we take $K = 5$. The first column $X_1 = 1$ and the other four covariates are generated randomly from a $N(0,1)$ distribution but are fixed throughout the simulation. The errors ϵ_i are generated from a $N(0, \sigma^2)$ distribution with $\sigma^2 = 1$. Five different β values are considered: $(2,0,0,4,0)'$, $(2,0,0,4,8)'$, $(2,9,0,4,8)'$, $(2,9,6,4,8)'$. These represent situations with generally strong signal to noise ratios. We also consider one other situation in which the true model is the full model but the signal to noise ratio is much smaller. This is the situation with $\beta = (1,2,3,2,3)'$ [see Rao (1999)]. All simulations were run 100 times. Two different sample sizes were considered: $n = 100$ and $n = 250$. We considered the fence based on (1.20) with $c = 1$ and that with $c = 1.1$, a slight increase, for each situation. Empirical probabilities of correct selection, underfitting (UF), and overfitting (OF) are reported in Table 2.1, where correct selection means that exactly those variables with the nonzero coefficients are selected; UF means that at least one variable with the nonzero coefficient is not included in the selected model; while OF means that the selected model includes all the variables with the nonzero coefficients, plus at least one variable with zero coefficient (Exercise 2.1). For comparison purposes, we also studied the performance of traditional methods, namely, Mallow's C_p (e.g., Sen and Srivastava (1990), sec. 11.2.4) and *BIC* in this simulation.

Summary: The fence method seems to have robust selection performance in most situations considered. In cases where the true model is small in dimension relative to the full model (take for example the case when $\beta' = (2,0,0,4,0)$), the fence method suffers some from overfitting. This can be mitigated by increasing c from 1 to 1.1. The overfitting prone-

Table 2.1 **Simulation Results: Regression Model Selection.** *Reported are probabilities of correct selection (underfitting, overfitting) as percentages estimated empirically from 100 simulations.*

True Model	n	σ	c_n (Fence only)	Fence	C_p	BIC
					% Correct (UF, OF)	
$\beta' = (2,0,0,4,0)$	100	1	1	69(0,31)	59(0,41)	89(0,11)
$\beta' = (2,0,0,4,0)$	100	1	1.1	73(0,27)	-	-
$\beta' = (2,0,0,4,0)$	250	1	1	71(0,29)	61(0,39)	94(0,6)
$\beta' = (2,0,0,4,0)$	250	1	1.1	87(0,13)	-	-
$\beta' = (2,0,0,4,8)$	100	1	1	79(0,21)	69(0,31)	92(0,8)
$\beta' = (2,0,0,4,8)$	100	1	1.1	95(0,5)	-	-
$\beta' = (2,0,0,4,8)$	250	1	1	83(0,17)	71(0,29)	95(0,5)
$\beta' = (2,0,0,4,8)$	250	1	1.1	92(0,8)	-	-
$\beta' = (2,9,0,4,8)$	100	1	1	93(0,7)	85(0,15)	97(0,3)
$\beta' = (2,9,0,4,8)$	100	1	1.1	94(0,6)	-	-
$\beta' = (2,9,0,4,8)$	250	1	1	97(0,3)	84(0,16)	98(0,2)
$\beta' = (2,9,0,4,8)$	250	1	1.1	99(0,1)	-	-
$\beta' = (2,9,6,4,8)$	100	1	1	100(0,0)	100(0,0)	100(0,0))
$\beta' = (2,9,6,4,8)$	100	1	1.1	100(0,0)	-	-
$\beta' = (2,9,6,4,8)$	250	1	1	100(0,0)	100(0,0)	100(0,0)
$\beta' = (2,9,6,4,8)$	250	1	1.1	100(0,0)	-	-
$\beta' = (1,2,3,2,3)$	100	1	1	100(0,0)	52(48,0)	35(65,0)
$\beta' = (1,2,3,2,3)$	100	1	1.1	100(0,0)	-	-
$\beta' = (1,2,3,2,3)$	250	1	1	100(0,0)	78(22,0)	69(31,0)
$\beta' = (1,2,3,2,3)$	250	1	1.1	100(0,0)	-	-

ness in these few situations is less than that found when using C_p but more than that found when using BIC. Selection performance in the other situations considered is typically solid for the fence method. Performance generally improves for all methods when the sample size increases as would be expected. The BIC, however, is known to be prone to underfitting. This is brought out by looking at the situation where $\beta' = (1,2,3,2,3)$. This is a case of weak signal but the optimal model being the full model. Here the fence method still shines having excellent performance with no over or underfitting empirically observed. Note how BIC underfits badly in this case and note, interestingly, that even C_p underfits here.

2.1.2 *Linear mixed models (clustered data)*

This subsection may be viewed as an extension of the previous one. We consider selection in the following simple mixed linear model considered by

Jiang and Rao (2003),
$$y_{ij} = \beta_0 + \beta_1 x_{ij} + \alpha_i + \epsilon_{ij} , \qquad (2.2)$$
$i = 1, \ldots, m$, $j = 1, \ldots, K$, where β_0, β_1 are unknown coefficients (the fixed effects). It is assumed that the random effects $\alpha_1, \ldots, \alpha_m$ are uncorrelated with mean 0 and variance σ^2. Furthermore, assume that the errors ϵ_{ij}'s have the following exchangeable correlation structure: Let $\epsilon_i = (\epsilon_{ij})_{1 \leq j \leq K}$. Then, $\text{cov}(\epsilon_i, \epsilon_{i'}) = 0$ if $i \neq i'$, and $\text{var}(\epsilon_i) = \tau^2\{(1 - \rho)I + \rho J\}$, where I is the identity matrix and J matrix of 1's. Finally, assume that the random effects are uncorrelated with the errors.

In the following simulation study, we consider a model similar to (2.2) except that more than one fixed covariates may be involved, i.e., $\beta_0 + \beta_1 x_{ij}$ is replaced by $x_{ij}'\beta$, where x_{ij} is a vector of covariates and β a vector of unknown regression coefficients. We examine by simulation the probability of correct selection and also the empirical UF and OF (see the previous subsection) probabilities respectively of various GIC's developed in Jiang and Rao (2003), which are similar to (1.5) for the current problem. Two GIC's with different choices of λ_n are considered: (1) $\lambda_n = 2$, which corresponds to the C_p method; and (2) $\lambda_n = \log n$, which corresponds to the BIC method, where $n = mK$. In the latter case we ignore the issue about the effective sample size discussed in §1.2.1. Nevertheless, the choice of (2) satisfies the conditions of Theorem 1 in Jiang and Rao (2003) for consistent model selection for the case of a single random effect factor in the true underlying model with bounded cluster size, which includes the current case. A total of 100 realizations of each simulation scenario were run. The first column of the full matrix of covariates, X, is 1_n, and the next four columns of X are generated randomly from $N(0, 1)$ distributions, then fixed throughout the simulation. Three true β values are considered: $(2, 0, 0, 4, 0)$, $(2, 9, 0, 4, 8)$ and $(1, 2, 3, 2, 3)$, which are the part of the situations previously studied in Subsection 2.1.1, with the first two representing situations with stronger signal to noise ratio as compared to the last one. However, unlike the previous subsection, the data are correlated in the current case (Exercise 2.2).

We consider the case where the correlated errors have varying degrees of exchangeable structure as described above, where four values of ρ are entertained: $0, 0.2, 0.5, 0.8$. The random effects and errors are simulated from the normal distributions with σ and τ both taken to be equal to 1. We set the number of clusters (m) to be 100 and the number of observations within a cluster to be fixed at $K = 5$. The negative log-likelihood measure (Example 1.3) is applied with $c = 1.1$ in (1.20) for all of the cases.

Table 2.2 **Simulation Results: Linear Mixed Model Selection.** *Reported are probabilities of correct selection (underfitting, overfitting) as percentages estimated empirically from 100
simulations. C_p and BIC results for models 1 and 2 were taken
from Jiang and Rao (2003).*

True Model	ρ	C_p	BIC	Fence (ML)
$\beta' = (2,0,0,4,0)$	0	64(0,36)	97(0,3)	94(0,6)
	0.2	57(0,43)	94(0,6)	91(0,9)
	0.5	58(0,42)	96(1,3)	86(0,14)
	0.8	61(0,39)	96(0,4)	72(0,28)
$\beta' = (2,9,0,4,8)$	0	87(0,13)	99(0,1)	100(0,0)
	0.2	87(0,13)	99(0,1)	100(0,0)
	0.5	80(0,20)	99(0,1)	99(0,1)
	0.8	78 (1,21)	96(1,3)	94(0,6)
$\beta' = (1,2,3,2,3)$	0	85(15,0)	81(19,0)	100(0,0)
	0.2	79(21,0)	73(27,0)	100(0,0)
	0.5	74(26,0)	64(36,0)	97(3,0)
	0.8	44(56,0)	26(74,0)	94(6,0)

Summary: The results are presented in Table 2.2. It is evident that
a similar picture to that presented in Table 2.1 arises. The fence method
seems to have robust selection performance in most situations considered.
In cases where the true model is small in dimension relative to the full model
$[\beta' = (2,0,0,4,0)]$, the fence method suffers some from overfitting. The
overfitting proneness in these few situations is less than that found when
using C_p but more than that found when using BIC. Selection performance
in the second situation with a larger true model with high signal is solid
for the fence method. However, in the last situation with the optimal
model being the full model with the weak signals, both BIC and C_p tend
to underfit. The fence method still shines having excellent performance
with comparatively little or no underfitting empirically observed (note that
overfitting is not possible in this situation since the true model is the full
model). The effect of increasing correlation in the errors (i.e., clustering)
seems to be to act as a means of reducing effective sample size for selection.
The end result is that as the correlation between observations within a
cluster increases, selection performance for all methods degrades somewhat.
The effect of the random effect is to essentially decrease the signal to noise
ratio relative to the situations considered in Subsection 2.1.1, where no
random effect was introduced and the errors were independent.

Table 2.3 **Simulation Results: Consistency.** *The columns for MVC and ML are probabilities of correct selection, reported as percentages estimated empirically from 100 simulations. The numbers in parentheses are the percentages of selection of the other two models in order of increasing index of the model.*

True Model	m	k	l	β_0	β_1	σ	c	MVC	ML
I	100	4	2	-.5	1	1	1	82(5,13)	94(3,3)
I	200	4	2	-.5	1	1	1.1	97(1,2)	99(0,1)
II	100	4	2	-.5	NA	1	1	87(4,9)	88(5,7)
II	200	4	2	-.5	NA	1	1.1	93(4,3)	98(2,0)
III	100	4	2	NA	NA	1	1	87(3,10)	91(2,7)
III	200	4	2	NA	NA	1	1.1	96(0,4)	91(1,8)

2.1.3   GLMMs (clustered data)

We consider the following simulated example of GLMM selection. Suppose that three models are being considered.

   <u>Model I</u>: Given the random effects $\alpha_1, \ldots, \alpha_m$, binary responses y_{ij}, $i = 1, \ldots, m$, $j = 1, \ldots, k$ are conditionally independent such that,

$$\text{logit}(p_{ij}) = \beta_0 + \beta_1 x_i + \alpha_i,$$

where $p_{ij} = \text{P}(y_{ij} = 1|\alpha)$; β_0, β_1 are fixed parameters; $x_i = 0$, $1 \leq i \leq [m/2]$ and $x_i = 1$, $[m/2] + 1 \leq i \leq m$ ($[x]$ represents the integer part of x). Furthermore, the random effects are independent and distributed as $N(0, \sigma^2)$.

   <u>Model II</u>: Same as Model I except that $\beta_1 = 0$.

   <u>Model III</u>: Same as Model I except that $\beta_0 = \beta_1 = 0$.

   We first study consistency of the MVC and ML model selection procedures in the situation where the data is generated from one of the candidate models (recall that ML model selection is based on the measure in Example 1.3). In other words, a true model belongs to the class of candidate models. Throughout the simulation studies, T was chosen as a block-diagonal matrix [see Example 1.6 (continued in Subsection 1.4.1)] with $T_i = T_1$, $1 \leq i \leq m$, where T_1 is a $k \times l$ matrix with $l = [k/2]$, whose entries are generated from a Uniform[0, 1] distribution, and then fixed throughout the simulations. The simulation results are summarized in Table 2.3, with each result based on 100 simulation runs.

   We next study robustness of the MVC and ML fence procedures in the case where no true model (with respect to ML) is included in the candidate models. We consider one such case, in which the binary responses y_{ij} are generated as follows. Suppose that $(X_1, \ldots, X_k)$ has a multivariate normal distribution such that $\text{E}(X_j) = \mu$, $\text{var}(X_j) = 1$, $1 \leq j \leq k$ and

Table 2.4 **Simulation Results: Robustness.** *The columns for MVC and ML are probabilities of correct selection, reported as percentages estimated empirically from 100 simulations. The numbers in parentheses are the percentages of selection of the other two models in order of increasing index of the model.* β_0^*, β_1^* *and* σ^* *are the matching parameters.*

True Model	m	k	l	β_0^*	β_1^*	σ^*	c	MVC	ML
A	100	4	2	-.5	1	1	1	83(7,10)	91(5,4)
A	200	4	2	-.5	1	1	1.1	97(2,1)	99(0,1)
B	100	4	2	-.5	NA	1	1	80(3,17)	91(4,5)
B	200	4	2	-.5	NA	1	1.1	95(3,2)	97(3,0)
C	100	4	2	NA	NA	1	1	83(8,9)	86(4,10)
C	200	4	2	NA	NA	1	1.1	91(1,8)	90(1,9)

$\text{cor}(X_s, X_t) = \rho$, $1 \leq s \neq t \leq k$. Then, let $Y_j = 1_{(X_j > 0)}$, $1 \leq j \leq k$. Denote the joint distribution of $(Y_1, \ldots, Y_k)$ by $\text{NB}(\mu, \rho)$ (here NB refers to "Normal-Bernoulli"; Exercise 2.3). We then generate the data such that $y_1, \ldots, y_m$ are independent, and the distribution of $y_i = (y_{ij})_{1 \leq j \leq k}$ follows one of the following models.

Model A: $y_i \sim \text{NB}(\mu_1, \rho_1)$, $i = 1, \ldots, [m/2]$, and $y_i \sim \text{NB}(\mu_2, \rho_2)$, $i = [m/2] + 1, \ldots, m$, where μ_j, ρ_j, $j = 1, 2$ are chosen to match the means, variances and covariances under Model I. Note that one can do so because the means, variances and covariances under Model I depend only on three parameters, while there are four parameters under Model A.

Model B: $y_i \sim \text{NB}(\mu, \rho)$, $i = 1, \ldots, m$, where μ and ρ are chosen to match the mean, variance and covariance under Model II. Note that, under Model II, the mean, variance and covariance depend on two parameters.

Model C: Same as Model B except that μ and ρ are chosen to match the mean, variance and covariance under Model III. Note that, under Model III, the mean is equal to $1/2$, the variance is $1/4$, while the covariance depends on a single parameter σ.

If the data is generated from Model A, Model I is a correct model with respect to MVC; similarly, if the data is generated from Model B, both Models I and II are correct with respect to MVC; and, if the data is generated from Model C, Models I - III are all correct in the sense of MVC. However, no model (I, II or III) is correct from an ML standpoint. The simulation results are summarized in Table 2.4, in which β_0^*, β_1^* and σ^* correspond to the parameters under the models in Table 2.3 with the matching mean(s), variance(s) and covariance(s). Each result is based on 100 simulation runs.

Summary: It is seen in Table 2.3 that the numbers increase as m increases (and c slowly increases), a good indication of consistency. With the exception of one case (III/200), ML outperforms MVC, which is not surprising. What is a bit of surprise is that ML also seems quite robust in the situation where the true model is not one of the candidate models (therefore the objective is to select a model among the candidates that is closest to the reality). In fact, Table 2.4 shows that even in the latter case, ML still outperforms MVC (with the exception of one case - again, III/200). However, one has to keep in mind that there are many ways that a model can be misspecified, and here we only considered one of them (which misspecifies a NB as a GLMM). Furthermore, MVC has computational advantage over ML, which is important in cases such as GLMM selection. Note that the computational burden usually increases with the sample size; on the other hand, the larger sample performance of MVC (i.e., $m = 200$) is quite close to that of ML. A compromise would be to use MVC in cases of large sample, and ML in cases of small or moderate sample. Alternatively, one may use MVC for an initial round of model selection to narrow down the number of candidate models, and ML for a final round of model selection.

2.1.4 *Gaussian model selection*

We consider the problem of selecting a Gaussian linear mixed model for non-clustered observations. There are three candidate models. These are:

Model I. $y_{ij} = \beta_0 + \beta_1 x_{ij} + u_i + v_j + e_{ij}$, $i = 1, \ldots, a$, $j = 1, \ldots, b$, where β_0 and β_1 are unknown coefficients, u_i, v_j are random effects, and e_{ij} is an error. It is assumed that u_i's, v_j's and e_{ij}'s are independent with $u_i \sim N(0, \sigma_1^2)$, $v_j \sim N(0, \sigma_2^2)$ and $e_{ij} \sim N(0, \sigma_0^2)$.

Model II. $y_{ij} = \beta_0 + u_i + v_j + e_{ij}$, where everything is the same as in Model I.

Model III. $y_{ij} = \beta_0 + \beta_1 x_{ij} + u_i + e_{ij}$, where everything is the same as in Model I.

In the simulation, the x_{ij}'s are generated from a Poisson(1) distribution and, once generated, fixed throughout the simulation.

We consider the fence ML model selection (Example 1.3), which seems to be the nature choice in this case. Note that this is a case of non-clustered data (Exercise 2.4). We consider the fence with $c = 1$ in (1.20). Four sample size configurations are considered: (i) $a = b = 10$; (ii) $a = 10, b = 20$; (iii) $a = 20, b = 10$; and (iv) $a = b = 20$. Note that the effective sample sizes here are a and b, not the product ab (see the discussion in §1.2.1), so these configurations correspond to situations of relatively small sample size.

For each sample size configuration, three cases are considered. In the

Table 2.5 **Gaussian Model Selection.** *Reported are probabilities of correct selection as percentages estimated empirically from 100 simulations.*

Optimal Model	$a = b = 10$	$a = 10, b = 20$	$a = 20, b = 10$	$a = b = 20$
Model I	34	92	85	97
Model II	97	80	79	96
Model III	92	98	98	99

first case, the data is generated under Model I with the following true parameters: $\beta_0 = 0.5$, $\beta_1 = 0.2$, $\sigma_j^2 = 1.0$, $j = 0, 1, 2$. In this case, Model I is the only true model and therefore the optimal model. In the second case, the data is generated under Model II with the following true parameters: $\beta_0 = 0.5$, $\sigma_j^2 = 1.0$, $j = 0, 1, 2$. In this case, Model I and Model II are both true models with Model II being the optimal model. In the third case, the data is generated under Model III with the following true parameters: $\beta_0 = 0.5$, $\beta_1 = 0.2$, $\sigma_j^2 = 1.0$, $j = 0, 2$. In this case, Model I and Model III are both true models with Model III being the optimal model.

For each combination of sample size configuration and case, one hundred simulations were run. Table 2.5 reports the percentages of simulations (out of the 100) in which the fence has selected the optimal model.

It is seen that, despite the relatively small sample size, the performance of the fence is satisfactory in all but one case. The exception occurs when $a = b = 10$ and data is generated from Model I. A closer look at this case reveals that all the misses went to Model II, which has the same random effect factors but no covariates (i.e., $\beta_1 = 0$). Some possible explanations are: (1) weak signal/noise ratio. Note that the true $\beta_1 = 0.2$, while all three variance components are equal to 1.0; and (2) small sample size. In this case, $s(M, \tilde{M})$ is obtained using the Gaussian formula derived in Subsection 1.4.2. However, all the variance components are involved in this formula, therefore have to be estimated. As mentioned, the effective sample size for estimating σ_1^2 is $a = 10$, and that for estimating σ_2^2 is $b = 10$. With such small sample sizes, the estimators of these variances are not expected to be accurate.

2.1.5 *AR model selection*

One of the traditional areas of applications of the information criteria is AR model selection. In this subsection, we apply the fence method to the problem of selecting the order of the AR, and compare performance of the fence with that of AIC and BIC.

The measure $Q(M)$ of in Example 1.4 is naturally used for the fence.

As for the evaluation of $s(M, \tilde{M})$, note that the computation of $Q(M)$ in this case is fairly straightforward. Thus, the following parametric bootstrap method Efron and Tibshirani (1993) may be used to obtain $s(M, \tilde{M})$. First note that, if $\tilde{M}$ is a true model and $b_{\tilde{M},j}$, $1 \leq j \leq p_{\tilde{M}}$ are the corresponding true coefficients, then we have $\mathrm{E}\{Q(\tilde{M})\} = \sum_{t=P+1}^{T} \mathrm{E}(\epsilon_t^2) = (T - P)\sigma^2$. Thus, an empirical method of moments [EMM; Jiang (2003)] estimator of σ^2 is given by $\hat{\sigma}^2 = Q(\tilde{M})/(T - P)$. We thus proceed as follows:
(i) Generate $y_1^*, \ldots, y_T^*$ from an $\mathrm{AR}(p_{\tilde{M}})$ model with coefficients $\hat{b}_{\tilde{M},j}$, $1 \leq j \leq p_{\tilde{M}}$ and $\sigma^2 = \hat{\sigma}^2$ given above;
(ii) compute $d^* = Q^*(M) - Q^*(\tilde{M})$, where $Q^*(M)$ $[Q^*(\tilde{M})]$ is computed the same way as $Q(M)$ $[Q(\tilde{M})]$ except using $y_1^*, \ldots, y_T^*$ instead of $y_1, \ldots, y_T$;
(iii) repeat (i), (ii) B times, thus obtain $d_1^*, \ldots, d_B^*$; let

$$s^2(M, \tilde{M}) = \frac{1}{B-1} \sum_{b=1}^{B} (d_b^* - \bar{d}^*)^2,$$

where $\bar{d}^* = B^{-1} \sum_{b=1}^{B} d_b^*$.

Five AR models are considered as candidate models, which correspond to orders $1, \ldots, 5$. The optimal model is an $\mathrm{AR}(2)$ model, which can be expressed as $y_t - 0.1y_{t-1} - 0.2y_{t-2} = \epsilon_t$, $t = 0, \pm 1, \pm 2, \ldots$. It follows that $\mathrm{AR}(p)$, $p = 2, 3, 4, 5$ are true models, while $\mathrm{AR}(1)$ is an incorrect model. For each simulated dataset, the order is selected by the fence, and then by AIC and BIC using the traditional procedures.

We consider the fence with $c = 1$ in (1.20). Two different sample sizes are considered: $T = 100$ and $T = 200$. In each case, the bootstrap sample size, B, matched the sample size T. Furthermore, three different distributions for the innovations ϵ_t are considered: (a) $N(0, 1)$; (b) normal mixture (NM); and (c) centralized exponential (CE), where (b) is the mixture of $N(-1, 9)$ with probability 0.8 and $N(4, 1)$ with probability 0.2; and (c) is the distribution of $X - 1$, where $X \sim \mathrm{Exponential}(1)$. Note that all distributions have mean 0 and variance 1 (Exercise 2.5). One hundred simulations were run for each combination of sample size and distribution. We report the simulated percentages of correct selection, mean and s.d. of the selected AR orders. The results are summarized in Table 2.6.

Summary: The results suggest that the fence performs very competitively with the AIC and BIC. In fact, the fence has the highest empirical percentage of correct selection in all cases; its simulated mean is closest to the true order (i.e., 2) in all cases with one tie with the AIC; and its simulated s.d. is smaller than that of the AIC in all cases with one tie. Another

Table 2.6 **AR Model Selection.** *Reported are probabilities of correct selection (as percentages), mean and s.d. of the selected AR order, all estimated empirically from 100 simulations. Here T is the sample size and B the (parametric) bootstrap sample size used in the fence.*

$T = B$	Method	Gaussian			NM			CE		
		%	mean	s.d.	%	mean	s.d.	%	mean	s.d.
100	Fence	45	2.0	1.1	48	1.9	1.0	42	1.8	1.0
	AIC	41	1.9	1.9	39	1.7	1.6	41	1.4	1.2
	BIC	35	1.4	0.5	34	1.4	0.5	31	1.3	0.5
200	Fence	67	2.0	0.7	77	2.1	0.7	72	2.2	1.0
	AIC	61	2.4	1.7	67	2.5	1.7	70	2.2	1.0
	BIC	57	1.6	0.5	69	1.8	0.5	67	1.7	0.5

notable pattern is that the reduction in s.d. as the sample size increases is more significant for the fence than for the AIC and BIC. In fact, the s.d.'s for BIC are almost unchanged going from 100 to 200, even though its % and mean have improved quite a bit. Finally, all three methods appear to be robust against non-normality of the innovations.

2.2 Real data examples

This section consists of four real-data examples.

2.2.1 *The salamander data*

The infamous salamander-mating data [McCullagh and Nelder (1989), sec. 14.5] and the associated GLMM have been extensively studied [e.g., Breslow and Clayton (1993), Lin and Breslow (1996), Booth and Hobert (1999), Jiang and Zhang (2001), and Torabi (2012)]. However, in most studies it has been assumed that a different group of animals (20 for each sex) are used in each mating experiment, although, in reality, the same group of animals were repeatedly used in two of the three experiments. The GLMMs used in these studies assumed that no further correlation among the data exists given the random effects. However, the responses in this case should be considered longitudinal, because repeated measures were collected from the same subjects (once in the summer and once in the fall). Therefore, serial correlation may still exist among the repeated responses given the female and male random effects. Alternatively, one could pool the responses from the two experiments involving the same group of animals, as suggested by McCullagh and Nelder (1989), sec. 4.1, so let $y_{ij\cdot} = y_{ij1} + y_{ij2}$, where y_{ij1} and y_{ij2} represent the responses from the summer and first fall experiments, respectively, that involved the same (ith) female and (jth) male. This avoids the issue of conditional independence, but brings in a new problem: The pooled response $y_{ij\cdot}$ may not be *binomial* given the random effects (Exercise 2.6).

In general, pooling the responses from the repeated measures over time will maintain conditional independence, but may destroy the (conditional) exponential family, another key assumption of GLMM. To address such concerns, Jiang and Zhang (2001) proposed the extended GLMM (see Example 1.7). The covariance matrix Σ depends on a vector ψ of variance components. A question is: how to select the function $h(\cdot)$, the fixed covariates (which are components of x_i) and the random effect factors (which correspond to subvectors of α). In other words, we have a problem of selecting a model for the conditional means.

To answer the question, let β_M and ψ_M denote β and ψ under a candidate model, M, and $g_{M,i}(\beta_M, \psi_M) = \mathrm{E}\{h_M(x_i'\beta_M + z_i'\Sigma_M^{1/2}\xi)\}$, where h_M is the function h under M, Σ_M is the covariance matrix under M evaluated at ψ_M, and the expectation is taken with respect to $\xi \sim N(0, I_m)$ (which does not depend on M). Here m is the dimension of α. The measure Q of (1.14) is used for the fence with $c = 1$ in (1.20). We use the $s(M, \tilde{M})$ derived in Subsection 1.4.4. Note that the observed-variance approach in Example 1.7 only applies to cluster data, which is not the case here. The evaluation of $g_{M,i}$ and $g_{M,i,j}$ requires integration with respect to the distribution of $\zeta_i = z_i'\hat{\Sigma}_M\xi$ and the joint distribution of (ζ_i, ζ_j). Note that, although the dimension of ξ is m, the dimensions of ζ_i is one, and that of (ζ_i, ζ_j) is two. Such integrals may be evaluated using numerical integration or Monte-Carlo methods.

Natural considerations for the link function include the *logit* and *probit*, which correspond to Models I and III below. Let Y_{ij1} be the observed proportion of successful matings between the ith female and jth male in the two experiments that involve the same female and male (i.e., the summer and first fall experiments). Let Y_{ij2} be the indicator of successful mating between the ith female and jth male in the last (fall) experiment involving a new set of animals. We assume that given the random effects, $u_{k,i}$, $v_{k,j}$, $k = 1, 2$, $i, j = 1, \ldots, 20$, which are independent and normally distributed with mean 0 and variances σ^2 and τ^2, respectively, the responses Y_{ijk}, $(i, j) \in P$, $k = 1, 2$ are conditionally independent, where P represents the set of pairs (i, j) determined by the design, which is partially crossed; u and v represent the female and male, respectively; $1, \ldots, 10$ correspond to RB (rough butt), and $11, \ldots, 20$ correspond to WS (whiteside). Furthermore, we consider the following models for the conditional means:

Model I: $E(Y_{ijk}|u,v) = h_1(\beta_0 + \beta_1 \text{WS}_f + \beta_2 \text{WS}_m + \beta_3 \text{WS}_f \times \text{WS}_m + u_{k,i} + v_{k,j})$, $(i,j) \in P$, $k = 1, 2$, where $h_1(x) = e^x/(1 + e^x)$; WS_f is an indicator for WS female (1 for WS and 0 for RB), WS_m is an indicator for WS male (1 for WS and 0 for RB) and $\text{WS}_f \times \text{WS}_m$ represents the interaction.

Model II: Same as Model I except dropping the interaction term.

Model III: Same as Model I with h_1 replaced by h_2, where $h_2(x) = \Phi(x)$, the cdf of $N(0,1)$.

Model IV: Same as Model III except dropping the interaction term.

The analysis has yielded the following values of $Q(M)$ for $M = $ I, II, III and IV: $39.5292, 44.3782, 39.5292, 41.6190$. Therefore, we have $\tilde{M} = $ I or III. If we use $\tilde{M} = $ I, then $s(M, \tilde{M}) = 1.7748$ for $M = $ II and $s(M, \tilde{M}) = 1.1525$ for $M = $ IV. Therefore, neither $M = $ II nor $M = $ IV fall within the fence. If we use $\tilde{M} = $ III, then $s(M, \tilde{M}) = 1.68$ for $M = $ II and $s(M, \tilde{M}) = 1.3795$ for $M = $ IV. Thus, once again, neither $M = $ II nor $M = $ IV are inside the fence. In conclusion, the fence method has selected both Model I and Model III (either one) as the optimal model. Interestingly, these are exactly the ones fitted by Jiang and Zhang (2001) using a different method, although the authors had not considered it a model selection problem. The eliminations of Model II and Model IV are consistent with many of the previous studies (e.g., Karim and Zeger (1992), Breslow and Clayton (1993), Lin and Breslow (1996)), which have found the interaction term significant, although the majority of these studies have focused on logit models.

2.2.2 *The diabetes data: The fence with LAR*

Efron *et al.* (2004) proposed the least angle regression and selection (LARS) as a less greedy alternative to traditional forward model selection methods in regression settings. LARS represents a computationally elegant algorithm for generating stagewise-like regression solutions. Constrained versions of LARS can produce sparse solutions in the form of the Lasso [Tibshirani (1996)] or forward stagewise linear regression. Each can be used for regression model selection when different candidate LARS models are compared using a goodness of fit measure.

The LARS procedure builds up regression estimates in successive steps, each step adding one covariate to the model, so that after M steps, just M of the $\hat{\beta}_j$'s are zero. Algorithmically, it can be described as follows [taken directly from Tibshirani (2005)]:

a. Start with all coefficients $\hat{\beta}_j = 0$.

b. Find the predictor x_j most correlated with y.

c. Increase the coefficient $\hat{\beta}_j$ in the direction of the sign of its correlation with y. Take residuals $r = y - \hat{y}$ along the way. Stop when some other predictor x_k has as much correlation with r as x_j has.

d. Increase $(\hat{\beta}_j, \hat{\beta}_k)$ in their joint least squares direction until some other predictor has as much correlation with r.

e. Continue until all predictors are in the model.

Efron *et al.* (2004) chose to use a version of the C_p to determine which LARS model is preferable (i.e., which minimized their C_p criterion). As noted in the discussion to their paper, while the C_p can give an unbiased estimate of prediction error for a given model assuming σ^2 is known, its use can lead to overfit models. We attempted to use the fence method in this situation. Our interest lies in selecting an optimal model size expressed as the degrees of freedom [Efron *et al.* (2004)]. These degrees of freedom more or less correspond to stepwise updates from the LARS algorithm. More specifically, we consider the same dataset used by Efron *et al.* (2004). The outcome of interest is a quantitative measure of progression of diabetes one year after baseline taken on $n = 442$ diabetes patients. This is to be modeled as a function of ten baseline variables that include age, sex, body mass index (bmi), average blood pressure (map), and six blood serum measurements. According to the C_p, an optimal model seems to be found at a value of $K = 8$ predictors. Applying the fence method, a natural choice for the Q is $Q(M) = C_p(M)$. On the other hand, the evaluation of $s(M, \tilde{M})$ is a bit tricky since a closed form expression of $s(M, \tilde{M})$ does not exist. However, we can, again, use the bootstrap [Efron and Tibshirani (1993)], as follows.

Because the observations are clustered, by (1.15), $s^2(M, \tilde{M})$ can approximated by $E_* \left[\sum_{i=1}^{m} (Q_i - \tilde{Q}_i)^2 - \sum_{i=1}^{m} \{E_*(Q_i) - E_*(\tilde{Q}_i)\}^2 \right]$, where E_* is expectation taken with respect to F_n–the empirical distribution over the observed data. This can be approximated by Monte Carlo resampling using the following algorithm:

1. Find $\tilde{M}$ such that $Q(\tilde{M}) = \min_{M \in \mathcal{M}} Q(M)$.

2. For each $M \in \mathcal{M}$ such that $|M| < |\tilde{M}|$, compute $s(M, \tilde{M})$ by a Monte Carlo resampling. This is done with the following steps, which is similar to the bootstrap approach in Subsection 2.1.5 except that the latter used parametric bootstrap:

2.1. Fix $\tilde{M}$ and $|M|$. Draw bootstrap samples by sampling with replacement from the data pairs $(x_i, y_i); i = 1, \ldots, n$.

2.2. For the bth bootstrap sample, find $Q^*(\tilde{M})$, which is $Q(\tilde{M})$ computed with the bootstrap data. Also find $Q^*(M)$ for the current M. Evaluate $d_b = Q^*(M) - Q^*(\tilde{M})$. Repeat for $b = 1, \ldots, B$ for some large B. Then, let $s^2(M, \tilde{M}) = (B-1)^{-1} \sum_{b=1}^{B} (d_b^* - \bar{d}^*)^2$ where $\bar{d}^* = \sum_b d_b^*/B$. Repeat for each $M \in \mathcal{M}$ such that $|M| < |\tilde{M}|$.

3. If $Q(M) \leq Q(\tilde{M}) + 1.1 s(M, \tilde{M})$, M belongs to $\tilde{\mathcal{M}}_-$, the set of "true" models with smaller dimension than $\tilde{M}$. The rest of the procedure is the same as the fence.

The fence method, with $c = 1.1$ in (1.20), produces a model with $K = 5$ predictors. These are sex, bmi, map, ldl, hdl and ltg (last three are blood serum measurements) only. Interestingly, the fence method has excluded tc, tch and glu which were included in the minimal C_p model for the LARS. Note that the excluded are all blood serum measurements, while three other blood serum measurements made themselves to the fence-selected model. Obviously, we do not know the true model for the diabetes study. However, the model selected by the fence method actually coincides exactly with the model chosen by the spike and slab variable selection method of Ishwaran and Rao (2005) applied to this data. This method is known to produce good solutions for regression model selection situations.

2.2.3 Analysis of Gc genotype data

Human group-specific component (Gc) is the plasma transport protein for Vitamin D. Polymorphic electrophoretic variants of Gc are found in all human populations. Daiger *et al.* (1984) presented data involving a series of monozygotic (MZ) and dizygotic (DZ) twins of known Gc genotypes in order to determine the heritability of quantitative variation in Gc. These included 31 MZ twin pairs, 13 DZ twin pairs, and 45 unrelated controls. For each individual, the concentration of Gc was available along with additional information about the sex, age, and Gc genotype of the individual. The genotypes are distinguishable at the Gc structural locus, and are classified as 1–1, 1–2, and 2–2.

Lange (2002) considered three statistical models for the Gc genotype data. Let y_{ij} represent the Gc concentration measured for the jth person who is one of the ith identical twin pair, $i = 1, \ldots, 31$, $j = 1, 2$. Furthermore, let y_{ij} represent the Gc concentration measured for the jth person who is one of the $(i - 31)$th fraternal twin pairs, $i = 32, \ldots, 44$, $j = 1, 2$. Finally, Let y_i represent the Gc concentration for the $(i - 44)$th person among the unrelated controls, $i = 45, \ldots, 89$. Then, the first model, Model

I, can be expressed as

$$y_{ij} = \mu_{1-1}1_{(g_{ij}=1-1)} + \mu_{1-2}1_{(g_{ij}=1-2)} + \mu_{2-2}1_{(g_{ij}=2-2)}$$
$$+ \mu_{\text{male}}1_{(s_{ij}=\text{male})} + \mu_{\text{age}}a_{ij} + \epsilon_{ij}, \qquad i = 1, \ldots, 44, \qquad j = 1, 2,$$

where g_{ij}, s_{ij}, and a_{ij} represent the genotype, sex, and age of the jth person in the i twin pair (identical or fraternal), and ϵ_{ij} is an error that is further specified later. If we let x_{ij} denote the vector whose components are $1_{(g_{ij}=1-1)}$, $1_{(g_{ij}=1-2)}$, $1_{(g_{ij}=2-2)}$, $1_{(s_{ij}=\text{male})}$, and a_{ij}, and β denote the vector whose components are μ_{1-1}, μ_{1-2}, μ_{2-2}, μ_{male} and μ_{age}, then the model can be expressed as

$$y_{ij} = x'_{ij}\beta + \epsilon_{ij}, \qquad i = 1, \ldots, 44, \qquad j = 1, 2. \qquad (2.3)$$

Similarly, we have

$$y_i = \mu_{1-1}1_{(g_i=1-1)} + \mu_{1-2}1_{(g_i=1-2)} + \mu_{2-2}1_{(g_i=2-2)}$$
$$+ \mu_{\text{male}}1_{(s_i=\text{male})} + \mu_{\text{age}}a_i + \epsilon_i, \qquad i = 45, \ldots, 89,$$

where g_i, s_i, and a_i are the genotype, sex, and age of the $(i-44)$th person in the unrelated control group, and ϵ_i is an error that is further specified. Let x_i denote the vector whose components are $1_{(g_i=1-1)}$, $1_{(g_i=1-2)}$, $1_{(g_i=2-2)}$, $1_{(s_i=\text{male})}$, and a_i, and β be the same as above; then we have

$$y_i = x'_i\beta + \epsilon_i, \qquad i = 45, \ldots, 89. \qquad (2.4)$$

We now specify the distributions for the errors. Let $\epsilon_i = (\epsilon_{i1}, \epsilon_{i2})'$, $i = 1, \ldots, 44$. We assume that ϵ_i, $i = 1, \ldots, 89$ are independent. Furthermore, we assume that

$$\epsilon_i \sim N\left(\begin{bmatrix} 0 \\ 0 \end{bmatrix}, \sigma_{\text{tot}}^2 \begin{bmatrix} 1 & \rho_{\text{ident}} \\ \rho_{\text{ident}} & 1 \end{bmatrix}\right), \qquad i = 1, \ldots, 31,$$

where σ_{tot}^2 is the unknown total variance, and ρ_{ident} the unknown correlation coefficient between identical twins. Similarly, we assume

$$\epsilon_i \sim N\left(\begin{bmatrix} 0 \\ 0 \end{bmatrix}, \sigma_{\text{tot}}^2 \begin{bmatrix} 1 & \rho_{\text{frat}} \\ \rho_{\text{frat}} & 1 \end{bmatrix}\right), \qquad i = 32, \ldots, 44, \qquad (2.5)$$

where ρ_{frat} is the unknown correlation coefficient between fraternal twins. Finally, we assume that

$$\epsilon_i \sim N(0, \sigma_{\text{tot}}^2), \qquad i = 45, \ldots, 89.$$

The second model, Model II, is the same as Model I except under the constraint $\rho_{\text{frat}} = \rho_{\text{ident}}/2$; that is, in (2.5) ρ_{frat} is replaced by $\rho_{\text{ident}}/2$.

The third model, Model III, is the same as Model I except under the constraints $\mu_{1-1} = \mu_{1-2} = \mu_{2-2}$; that is, in (2.3) and (2.4) we have

$$x'_{ij}\beta = \mu + \mu_{\text{male}}1_{(s_{ij}=\text{male})} + \mu_{\text{age}}a_{ij} + \epsilon_{ij},$$
$$x'_i\beta = \mu + \mu_{\text{male}}1_{(s_i=\text{male})} + \mu_{\text{age}}a_i + \epsilon_i.$$

Thus, under Model I, the parameters are

$$\theta_{\text{I}} = (\mu_{1-1}, \mu_{1-2}, \mu_{2-2}, \mu_{\text{male}}, \mu_{\text{age}}, \sigma^2_{\text{tot}}, \rho_{\text{ident}}, \rho_{\text{frat}})'$$

(8-dimensional); under Model II, the parameters are

$$\theta_{\text{II}} = (\mu_{1-1}, \mu_{1-2}, \mu_{2-2}, \mu_{\text{male}}, \mu_{\text{age}}, \sigma^2_{\text{tot}}, \rho_{\text{ident}})'$$

(7-dimensional); and, under Model III, the parameters are

$$\theta_{\text{III}} = (\mu, \mu_{\text{male}}, \mu_{\text{age}}, \sigma^2_{\text{tot}}, \rho_{\text{ident}}, \rho_{\text{frat}})'$$

(6-dimensional).

It is clear that all three models are Gaussian mixed models (Exercise 2.7), which are special cases of GLMMs. We apply the fence method, with the ML model selection (Exercise 2.8), and $c = 1$ in (1.20), to this dataset to select an optimal model from the candidate models. Note that, because Models II and III are submodels of Model I (in other words, Model I is the full model), we may take $\tilde{M}$ as Model I. The analysis resulted in the following values for $Q(M)$: $Q(\text{I}) = 337.777$, $Q(\text{II}) = 338.320$, and $Q(\text{III}) = 352.471$. Furthermore, we obtained $s(\text{II}, \text{I}) = 1.367$ and $s\text{III}, \text{I}) = 4.899$. Thus, Model II is in the fence while Model III is not. In conclusion, the analysis has selected Model II as the optimal model. This result is consistent with the finding of Lange (2002), who indicated that a "likelihood ratio test shows that there is virtually no evidence against the assumption $\rho_{\text{frat}} = \rho_{\text{ident}}/2$."

2.2.4 *Prenatal care for pregnancy*

Finally, we consider a real-data example as a demonstration of the F-B fence, introduced in Section 1.5. Rodriguez and Goldman (2001) considered a dataset from a survey conducted in Guatemala regarding the use of modern prenatal care for pregnancies where some form of care was used [Pebley *et al.* (1996)]. While Rodriguez and Goldman (2001) focused on assessing the performance of the approximation method they developed in fitting a three-level variance component logistic model, we consider applying the fence method in selection of the fixed covariates in the variance-component logistic model. The models are described as follows.

Suppose that given the random effects at community levels u_i, $1 \leq i \leq m$ and random effects at family levels v_{ij}, $1 \leq i \leq m$, $1 \leq j \leq n_i$, binary responses y_{ijk}, $1 \leq i \leq m$, $1 \leq j \leq n_i$, $1 \leq k \leq n_{ij}$ are conditionally independent with $\pi_{ijk} = \mathrm{E}(y_{ijk}|u,v) = \mathrm{P}(y_{ijk} = 1|u,v)$. Furthermore, suppose that the random effects are independent with $u_i \sim N(0, \sigma^2)$ and $v_{ij} \sim N(0, \tau^2)$. The following models for the conditional means are considered such that under model M, $\mathrm{logit}(\pi_{ijk}) = X'_{M,ijk}\beta_M + u_i + v_{ij}$, where $X_{M,ijk}$ is a subvector of the full set of fixed covariates and β_M the corresponding vector of regression coefficeints (Exercise 2.9).

Let $\psi = (\sigma^2, \tau^2)'$. The vector of parameters under model M is $\theta_M = (\beta'_M, \psi')'$. Consider the measure

$$Q(M) = \sum_{i=1}^{m} \sum_{j=1}^{n_i} \sum_{k=1}^{n_{ij}} \{y_{ijk} - g_{M,ijk}(\theta_M)\}^2, \qquad (2.6)$$

where $g_{M,ijk}(\theta_M) = \mathrm{E}\{h(X'_{M,ijk}\beta_M + u_i + v_{ij})\}$ and $h(x) = e^x/(1 + e^x)$. Note that (2.6) is a special case of (1.14). Furthermore, we obtain $s(M, \tilde{M})$ using the method developed in subsection 1.4.4. The expectations involved in $Q(M)$ are evaluated by numerical intergration. Since the number of candidate covariates considered is relatively large (21), to keep the computing time manageable, the F-B fence is applied with $c = 1$ in (1.20).

The data analysis has selected the following variables (in the order that they were selected in the forward procedure): There are some interesting differences between the fixed effects discovered by the fence versus those found by standard maximum likelihood analysis using a 5% significance level as reported in Rodriguez and Goldman (2001). First, Husband's education overall (primary or higher relative to the reference group of no education for the husband) was found to be an important predictor whereas Rodriguez and Goldman found that only Husband's secondary education was important. Our more uniform finding is also in line with the finding for Mother's education. The implication is that education of some kind is important for both the mother and husband to have. A similar kind of finding was observed for variables corresponding to husband's profession. We found that regardless of what type of agricultural employment the husband had, it was an important predictor overall. Rodriguez and Goldman report that only non-self employed agricultural jobs for the husband mattered. The fence method also uniquely found that watching television (daily or not) was an important predictor. This can be intuitively justified since it provides a medium for women to learn more about modern prenatal health care methods and thus make it more likely for them to choose to use such

Table 2.7 **Variables Selected by the F-B Fence.**

Proportion indigenous (1981)
Modern toilet in household
Husband's education secondaryor better
Husband's education primary
Television watched daily
Distance to nearest clinic
Mother's education primary
Television not watched daily
Mother's education secondary or better
Indigenous (no Spanish)
Indigenous (Spanish)
Mother age
Husband agriculture employee
Husband agriculture self-employee
Child age
Birth order 4-6
Husband's education missing

methods. Other findings were in line with those of Rodriguez and Goldman (2001).

2.3 Exercises

2.1. The following questions are egarding Subsection 2.1.1:

(a) Identify the model(s) of correct selection.

(b) Give an example of an UF model.

(c) Give an example of an OF model.

(d) True or false: An UF model must have a dimension less than the total number of nonzero coefficients. Please explain your answer.

(e) True or false: An OF model must have a dimention higher than the total number of nonzero coefficents. Again, please explain your answer.

2.2. Let $y = (y_i)_{1 \leq i \leq m}$, where $y_i = (y_{ij})_{1 \leq j \leq K}$. Note that y is an $n \times 1$ vector, where $n = mK$. Derive the covariance matrix of y, Var(y).

2.3. Consider the NB model introduced in Subsection 2.1.3.

(a) Show that, if $\rho \neq 0$, NB is not a special case of GLMM.

(b) Show that, if $\rho = 0$, NB is a special case of GLMM.

2.4. Consider the Gaussian mixed model in Subsection 2.1.4. Derive the covariance matrix of $y = (y_{ij})_{1 \leq i \leq a}$, where $y_i = (y_{ij})_{1 \leq j \leq b}$. Show that the covariance matrix is not block-diagonal. Compare the result with that of Exercise 2.2. What is the main difference?

2.5. Verify that the NM and CE distributions in Subsection 2.1.5 have mean 0 and variance 1.

2.6. Show, by giving an example, that the conditional distribution of $y_{ij\cdot} = y_{ij1} + y_{ij2}$ given the female random effect, u_i, and male random effect, v_j, may not be binomial, if there is a serial correlation between y_{ij1} and y_{ij2} given u_i, v_j.

2.7. This and the next exercises are regarding Subsection 2.2.3. Obtain expressions of the mean vector and covariance matrix of y, the vector of the Gc concentrations, under the three candidate models, I, II, and III.

2.8. Obtain expressions of $Q(M)$, where Q is the measure for ML model selection (Example 1.3), for $M = $ I, II, and III.

2.9. Show that the model proposed in Subsection 2.2.4 for the prenatal care data is a special case of GLMM. Are the data here clustered or non-clustered?

Chapter 3

Adaptive Fence

A question that remains to be answered would be how to choose the tuning constant for the fence, that is, c in (1.20). In this chapter, we offer an approach to answer the question. The idea is called *adaptive fence*, originally due to Jiang *et al.* (2008) in the context of mixed model selection. Since then, several variations, and simplification, of the idea have been developed, which we shall discuss in this chapter, and in subsequent chapters.

3.1 The adaptive fence

Recall that $\mathcal{M}$ denotes the set of candidate models. To be more specific, we assume that the minimum-dimension criterion is used in selecting the models within the fence. Furthermore, we assume that there is a full model, $M_{\mathrm{f}} \in \mathcal{M}$, hence $\tilde{M} = M_{\mathrm{f}}$ in (1.20), and that every model in $\mathcal{M} \setminus \{M_{\mathrm{f}}\}$ is a submodel of a model in $\mathcal{M}$ with one less parameter than M_{f}. Also, let M_* denote a model with minimum dimension among $M \in \mathcal{M}$. First note that, ideally, one wishes to select c that maximizes the probability of choosing the optimal model. Here, by definition, the optimal model is a true model that has the minimum dimension among all of the true models. This means that one wishes to choose c that maximizes

$$P = \mathrm{P}(M_c = M_{\mathrm{opt}}), \tag{3.1}$$

where M_{opt} represents the optimal model, and M_c is the model selected by the fence (1.20) with the given c. However, two things are unknown in (3.1): (i) under what distribution should the probability P be computed? and (ii) what is M_{opt}?

To solve problem (i), note that the assumptions above on $\mathcal{M}$ imply that M_{f} is a true model. Therefore, it is possible to bootstrap under M_{f}. For example, one may estimate the parameters under M_f, then use a model-based (or parametric) bootstrap to draw samples under M_f. This allows us to approximate the probability distribution P on the right side of (3.1).

To solve problem (ii), we use the idea of maximum likelihood. Namely, let $p^*(M) = \mathrm{P}^*(M_c = M)$, where $M \in \mathcal{M}$ and P^* denotes the empirical probability obtained by the bootstrapping. In other words, $p^*(M)$ is the sample proportion of times out of the total number of bootstrap samples that model M is selected by the fence method with the given c. Let $p^* = \max_{M \in \mathcal{M}} p^*(M)$. Note that p^* depends on c. The idea is to choose c that maximizes p^*. It should be kept in mind that the maximization is not without restriction. To see this, note that if $c = 0$ then $p^* = 1$ (because, when $c = 0$, the procedure always chooses M_f). Similarly, $p^* = 1$ for very large c, if M_* is unique (because, when c is large enough, the procedure always chooses M_*). Therefore, what one looks for is "a peak in the middle" of the plot of p^* against c. Figure 3.1 shows a "typical" shape of such a plot, where $c_n = c$.

Here is another look at the method. Typically, the optimal model is the model from which the data is generated, then this model should be the most likely given the data. Thus, given c, one is looking for the model (using the fence procedure) that is most supported by the data or, in other words, one that has the highest (posterior) probability. The latter is estimated by bootstrapping. Note that although the bootstrap samples are generated under M_f, they are almost the same as those generated under the optimal model. This is because the estimates corresponding to the zero parameters are expected to be close to zero, provided that the parameter estimators under M_f are consistent. [Note that in some special cases, a non-model based bootstrap algorithm can also be used. For instance, in the case of crossed random effects, Owen (2007) presents a pigeonhole bootstrap algorithm which can be used effectively.] One then pulls off the c that maximizes the (posterior) probability and this is the optimal choice, denoted by c^*. We call this procedure the *adaptive fence*.

A few issues arise regarding the implementation of the adaptive fence. Quite often the search for the peak in the middle finds multiple peaks (see Figure 3.1). In such cases one strategy is to choose c corresponding to the highest peak. Although the strategy is supported by our theoretical result (see Chapter 9), that is, asymptotically, the highest peak corresponds to the optimal model, empirical studies show that the highest peak is not necessarily the best in a finite-sample situation. For example, Nguyen and Jiang (2012) finds that the "first significant peak" may be a better strategy

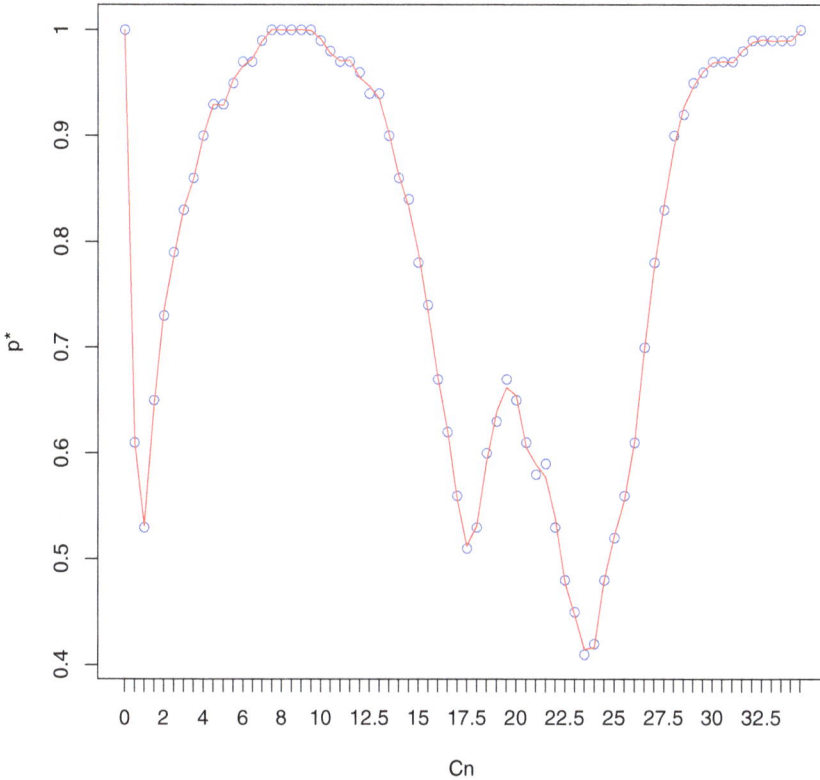

Fig. 3.1   *A plot of p^* against $c_n = c$*

in situations of moderate sample size, or relatively weak signal. Müller *et al.* (2013) finds a simple rule that was "surprisingly successful" in identifying the optimal model. Occasionally, the peak in the middle is "flat", that is, the value of p^* is constant over an interval. In such a case, a simple choice would be the middle point of the interval (or the median of a grid of values of c over which p^* is a constant). However, empirical studies may suggest that, for example, the lower bound of the interval (or the minimum of the c values at which p^* is the constant) may be a better choice.

There are also some technical issues regarding the situation when the optimal model is either M_f, or M_*. Nevertheless, these issues are mainly of theoretical interest. In most cases of variable selection there are a set of candidate variables and only some of them are important. This means that the optimal model is neither the full model nor the minimum model.

Therefore, we refer the (technical) details on how to handle these extreme cases to Jiang *et al.* (2008).

How does the idea work? We illustrate with two simulated examples.

Example 3.1: (The Fay-Herriot model) Fay and Herriot (1979) proposed the following model to estimate the per-capita income of small places with population less than 1,000. The model has since become popular in small area estimation [Rao (2003)]. The model can expressed as $y_i = x_i'\beta + v_i + e_i$, $i = 1, \ldots, n$, where x_i is a vector of known covariates, β is a vector of unknown regression coefficients, v_i's are area-specific random effects and e_i's represent sampling errors. It is assumed that v_i, e_i are independent with $v_i \sim N(0, A)$ and $e_i \sim N(0, D_i)$. The variance A is unknown, but the sampling variances D_i's are assumed known (Exercise 3.1).

Let $X = (x_i')_{1 \leq i \leq n}$, so that the model can be expressed as $y = X\beta + v + e$, where $y = (y_i)_{1 \leq i \leq n}$, $v = (v_i)_{1 \leq i \leq n}$ and $e = (e_i)_{1 \leq i \leq n}$. The first column of X is assumed to be 1_n which corresponds to the intercept. The rest of the columns of X are to be selected from a set of candidate covariate vectors $X_2, \ldots, X_K$, which include the true covariate vectors. In the simulation we let $D_i = 1$, $1 \leq i \leq n$.

We consider the fence ML model selection (Example 1.3). It is easy to show that, in this case, $Q(M) = (n/2)\{1 + \log(2\pi) + \log(|P_{X^\perp}y|^2/n)\}$, where $P_{X^\perp} = I_n - P_X$ and $P_X = X(X'X)^{-1}X'$. We assume for simplicity that X is of full rank. Then, we have $Q(M) - Q(M_f) = (n/2)\log(|P_{X^\perp}y|^2/|P_{X_f^\perp}y|^2)$. Furthermore, it can be shown that, when M is a true model, we have $Q(M) - Q(M_f) = (n/2)\log\{1 + (K - p)(n - K - 1)^{-1}F\}$, where $p + 1$ is the number of columns of X, and $F \sim F_{K-p,n-K-1}$ (Exercise 3.2). Therefore, $s(M, M_f)$ is completely known given $|M|$ and can be evaluated accurately (e.g., by numerical integration; Exercise 3.3).

We carry out a simulation study to evaluate the performance of the adaptive method. We consider a (relatively) small sample situation with $n = 30$. With $K = 5$, $X_2, \ldots, X_5$ were generated from the $N(0, 1)$ distribution, and then fixed throughout the simulations. The candidate models include all possible models with at least an intercept (thus there are $2^4 = 16$ candidate models). We consider five cases in which the data y is generated from the model $y = \sum_{j=1}^{5} \beta_j X_j + v + e$, where $\beta' = (\beta_1, \ldots, \beta_5) = (1, 0, 0, 0, 0), (1, 2, 0, 0, 0), (1, 2, 3, 0, 0), (1, 2, 3, 2, 0)$ and $(1, 2, 3, 2, 3)$, denoted by Model 1, 2, 3, 4, 5, respectively. The true value of A is 1 in all cases. The number of bootstrap samples for the evaluation of the p^*'s is 100.

In addition to the adaptive method, we consider five different (non-

Table 3.1 **Fence with Different Choice of c in the F-H Model.**

Optimal Model	1	2	3	4	5
Adaptive c	100	100	100	99	100
$c = \log\log(n)$	52	63	70	83	100
$c = \log(n)$	96	98	99	96	100
$c = \sqrt{n}$	100	100	100	100	100
$c = n/\log(n)$	100	91	95	90	100
$c = n/\log\log(n)$	100	0	0	0	6

adaptive) c's, which satisfy the consistency requirements given in Theorem 9.1. Note that, typically, c depends on the sample size, n, and therefore is denoted by c_n when studying the asymptotic behaviors. The consistency requirements reduce to $c_n \to \infty$ and $c_n/n \to 0$ in this case. The non-adaptive c's These are $c = \log\log(n)$, $\log(n)$, $\sqrt{n}$, $n/\log(n)$ and $n/\log\log(n)$. Reported in Table 3.1 are percentage of times, out of the 100 simulations, that the optimal model was selected by each method.

It seems that the performance of the fence with $c = \log(n)$, $\sqrt{n}$ or $n/\log(n)$ is fairly close to that of the adaptive fence. In any particular situation one might get lucky to find a good c value by chance, but one cannot be lucky all the time. Regardless, the adaptive fence always seems to pick up the optimal value, or something close to the optimal value of c in terms of the finite-sample performance.

Example 3.2: Let us consider one of the earlier examples, that is, the linear mixed model example considered in Subsection 2.1.2. This time, we include the adaptive fence (AF) in the comparison. Table 3.2 presents the results, where the results for C_p, BIC and Fence ($c = 1.1$) are copied from Table 2.2. As can be seen, the adaptive fence has perfect (100%) probability of correct selection in all cases considered. The example clearly demonstrates the effectiveness of the adaptive fence method.

It should be noted that the gain in the performance via the adaptive fence is at a computational cost, mainly due to the bootstrapping. To have some idea about the computational cost, consider, for example, the simulation study for the Fay-Herriot model (Example 3.1). The simulations were run on a Pentium Dual Core CPU 3.2GHz, memory 4GB, Hard drive 500 GB. The times it took to run the first simulation of for the adaptive fence under Models 1–5 were 1.7 sec., 3.0 sec., 4.1 sec., 4.4 sec. and 5.3 sec., respectively. The computational burden could become a bigger issue in high-dimensional selection problems. We shall address the computational issue in subsequent section and chapters.

Table 3.2 **Simulation Results: Linear Mixed Model Selection.** *Reported are probabilities of correct selection (underfitting, overfitting) as percentages estimated empirically from 100 realizations of the simulation. C_p and BIC results for models 1 and 2 were taken from Jiang and Rao (2003).*

True Model	ρ	C_p	BIC	Fence ($c = 1.1$)	AF
$\beta' = (2, 0, 0, 4, 0)$	0	64(0,36)	97(0,3)	94(0,6)	100(0,0)
	0.2	57(0,43)	94(0,6)	91(0,9)	100(0,0)
	0.5	58(0,42)	96(1,3)	86(0,14)	100(0,0)
	0.8	61(0,39)	96(0,4)	72(0,28)	100(0,0)
$\beta' = (2, 9, 0, 4, 8)$	0	87(0,13)	99(0,1)	100(0,0)	100(0,0)
	0.2	87(0,13)	99(0,1)	100(0,0)	100(0,0)
	0.5	80(0,20)	99(0,1)	99(0,1)	100(0,0)
	0.8	78 (1,21)	96(1,3)	94(0,6)	100(0,0)
$\beta' = (1, 2, 3, 2, 3)$	0	85(15,0)	81(19,0)	100(0,0)	100(0,0)
	0.2	79(21,0)	73(27,0)	100(0,0)	100(0,0)
	0.5	74(26,0)	64(36,0)	97(3,0)	100(0,0)
	0.8	44(56,0)	26(74,0)	94(6,0)	100(0,0)

3.2 Simplified adaptive fence

In the previous section, we showed that the adaptive fence may have outstanding finite-sample performance compared to the information criteria and non-adaptive fence (i.e., fence with the fixed cut-off c). On the other hand, there is a significant increase in the computational cost for the adaptive fence compared to the non-adaptive one. The computational burden can be a major factor in high-dimensional selection problems, for example, regression variable selection when there is a large number of candidate variables to select from. In this section, we consider an approach motivated by the adaptive fence to reduce the computational burden.

Note that the calculation of $Q(M)$ is often fairly straightforward. For example, in many cases $Q(M)$ can be chosen as the negative log-likelihood, or residual sum of squares. On the other hand, the computation of $s(M, \tilde{M})$ can be quite challenging. Sometimes, even if an expression can be obtained for $s(M, \tilde{M})$, its accuracy as an estimate of the standard deviation cannot be guaranteed in a finite sample situation. For such a reason, this step of the fence method has complicated its applications. Below we propose a simplified procedure that avoids the calculation of $s(M, \tilde{M})$.

We assume that $\mathcal{M}$ contains a full model, M_f, of which each candidate model is a submodel. Note that this is not a serious constraint because usually one can add a full model to $\mathcal{M}$, if it is not already included. For example, for selecting the fixed covariates one may include a model that contains all the candidate covariates, if such a model is not already under consideration. It follws that $\tilde{M} = M_f$. To come up with the new procedure, we absorb the term $s(M, \tilde{M})$ on the right side of (1.20) into the constant c, which is to be determined adaptively in a similar way as in the previous section. In other words, we let the adaptive tuning constant take care the product $c \cdot s(M, \tilde{M})$ in the fence inequality (1.20). Under this simplified procedure, a model M is in the fence if

$$Q(M) - Q(M_f) \leq c, \qquad (3.2)$$

where c is chosen adaptively as before. For the sake of completeness, we repeat essentially the same procedure as follows: For each $M \in \mathcal{M}$, let $p^*(M) = \mathrm{P}^*(M_c = M)$ be the empirical probability of selection for M, where M_c denotes the model selected by the fence procedure based on (3.2) with the given c, and P^* is obtained by bootstrapping under M_f. For example, under a parametric model one can estimate the model parameters under M_f and then use a parametric bootstrap to draw samples under M_f. Suppose that B samples are drawn, then $p^*(M)$ is simply the sample proportion (out of a total of B samples) that M is selected by the fence procedure corresponding to (3.2) with the given c. Let $p^* = \max_{M \in \mathcal{M}} p^*(M)$. Note that p^* depends on c. Let c^* be the c that maximizes p^* and this is our choice. We call this procedure simplified adaptive fence (SAF; Jiang *et al.* (2009)).

Another adjustment is also considered. Notice that p^* is a sample proportion based on the bootstrap samples. We may construct a (large sample) 95% confidence lower bound for the (bootstrap) population proportion,

$$p^* - 1.96\sqrt{\frac{p^*(1 - p^*)}{B}} \qquad (3.3)$$

where B is the bootstrap sample size. When selecting c that maximizes p^* we take (3.3) into account. More specifically, suppose that there are two peaks in the plot of p^* against c located at c_1 and c_2 such that $c_1 < c_2$. Let p_1^* and p_2^* be the heights of the peaks corresponding to c_1 and c_2. As long as p_1^* is greater than the confidence lower bound at p_2^*, that is, (3) with $p^* = p_2^*$, we choose c_1 over c_2. Clearly, the selection is in favor of smaller c in order to be more conservative. In other words, here we are more concerned with underfitting than overfitting. One reason is that, sometimes, the current selection can be used as a preliminary selection, or screening, for selection in the next round. If so, then by being conservative, we are able to at least keep the true variables among the candidates, as opposed to eliminating them, so that they can be picked up in the next round.

We illustrate the SAF using a simulated example.

Example 3.3: We consider the following linear mixed model, also known as the nested error regression (NER) model [Battese *et al.* (1988)]:

$$y_{ij} = x'_{ij}\beta + v_i + e_{ij}, \quad i = 1, \ldots, m, \ j = 1, \ldots, n_i, \quad (3.4)$$

where y_{ij} is the response from the jth sample unit in the ith small area (e.g., a county); x_{ij} is a vector of auxiliary data that is potentially associated with y_{ij}; β is a vector of unknown regression coefficients; v_i is an area-specific random effect; and e_{ij} is the sampling error. It is assumed that the v_i's and e_{ij}'s are independent such that $v_i \sim N(0, \sigma_v^2)$ and $e_{ij} \sim N(0, \sigma_e^2)$.

In the simulation study, the number of areas, m, is chosen as either 10 or 15. The n_i's are generated from a Poisson(3) distribution, then fixed throughout the simulations. The v_i and e_{ij} are both generated from the $N(0, 1)$ distribution. The components of the x_{ij} are to be selected from x_{ijk}, $k = 0, 1, \ldots, 5$, where $x_{ij0} = 1$; x_{ij1} and x_{ij2} are generated independently from $N(0, 1)$ and then fixed throughout; $x_{ij3} = x_{ij1}^2$, $x_{ij4} = x_{ij2}^2$, and $x_{ij5} = x_{ij1}x_{ij2}$. The model is chosen to imitate the one proposed by Battese *et al.* (1988) for the Iowa crops data. See Chapter 6 for more details.

The simulated data are generated under two models: I. the model that involves the linear terms only, i.e., x_{kij}, $k = 0, 1, 2$, with all the regression coefficients equal to 1; II. the model that involves both the linear and the quadratic terms, i.e., x_{kij}, $k = 0, \ldots, 5$, with all the regression coefficients equal to 1. We study the performance of the adaptive fence procedure introduced in Section 2 with or without using the confidence lower bound (3.3). Here $Q(M)$ is chosen as the negative log-likelihood (Example 1.3). The bootstrap sample size B is chosen as 100. The results based on 100 simulations are reported in Table 3.3 which present the empirical probabilities of correct model selection. The results show that even in these cases of fairly small m, the performance of the adaptive fence is quite satisfactory. Note that for Model I, the method does not behave quite as well for $m = 10$ as for Model II, but that this quickly disappears when $m = 15$.

The method that makes use of the confidence lower bound seems to perform better in the smaller m case but for the case of larger m the two methods are indistinguishable.

3.3 Statistical models for human genetics

As an application of the SAF, we consider a problem regarding quantitative trait loci (QTL) mapping in human genetics. Random mating is

Table 3.3 **An Illustration of SAF.** *Reported are empirical probabilities, in terms of percentage, based on 100 simulations that the optimal model is selected, where c.l.b. stands for confidence lower bound.*

Optimal Model	# of Clusters, m	Highest Peak	c.l.b.
Model I	10	82	87
Model I	15	99	99
Model II	10	98	99
Model II	15	100	100

often considered as a model in the latter field, assuming that genotypes are in steady-state conditions gradually. Because humans are not subject to breeding experiments, it is more difficult to study the correlation between the phenotypes and genetic markers. Therefore, family-based studies have become essential for studying human disease, particularly the sib pair studies. This is because the relatives are frequently more similar in phenotypic characteristics than non-relatives, presumably because they share more genetic material in common. Thus, it seems plausible to hypothesize that, for example, two relatives have the disease because they both inherited one or more disease-predisposing alleles from a common ancestor. More specifically, two relatives are said to have inherited an allele identical by descent (IBD) at a given locus, if they have inherited the same allele from a common ancestor. Therefore, we shall restrict our attention sib-pair studies.

Let $V_{n \times p}$ represents IBD information, where n is the number of sib pairs, and p is the total number of markers that have been genotyped, and $V_{ij} = V(i, j)$ is the number of alleles inherited identity by descent of the ith sib-pair at the jth marker. Thus, we have $P(\nu_{ij} = 0) = 1/4$, $P(\nu_{ij} = 1) = 1/2$, $P(\nu_{ij} = 2) = 1/4$. It follows that $E(\nu_{ij}) = 1$. Therefore, suppose that there is a sample of n sib pairs who share a trait (e.g., affected sib pairs by a certain disease). If a common marker is unlinked to gene(s) that contribute to the variation of the phenotype, one expects to see one allele inherited IBD for each sib pair on average; in other words, we expect to see n alleles inherited IBD for this sample of n sib pairs in total. On the other hand, if the marker is tightly linked to a gene(s) attributed to the phenotypic difference, we expect to see more than n alleles shared IBD in this sample. Furthermore, in sib pairs, the evidence for linkage of quantitative traits is that the conditional and unconditional covariance of the phenotype given IBD status at a QTL are different.

Although many simple traits have been mapped successfully with family-based study design, mapping genes in complex traits is still very challenging due to the small signal to noise ratio at a linked marker, and limited sample size. While several approaches have arisen in efforts to overcome such limitations, linkage analysis of quantitative traits by the use of sib pairs remains an important tool for genetic dissection of complex disorders in human. A regression-based linkage analysis was proposed by Haseman and Elston (1972) to detect QTL by linkage to a marker based on sib-pair data. In the latter method, squared sib pair trait difference is regressed on their estimated proportion of marker alleles that are shared IBD. The method has the advantages of simplicity and robustness Allison *et al.* (2000). However, as noted by Wright (1997), the use of squared difference does not capture all the information, since the trait data for a sib pair is bivariate in nature. Bivariate regression in which squared difference and squared sum in phenotypic values of sib pair being regressed on IBD was then proposed Drigalenko (1998), which has been shown to have power equivalent to that of the variance-components models for sib pairs cases under the normality assumption. We now introduce a model for QTL mapping using the sib-pairs.

3.3.1   *A model for QTL mapping*

Consider an additive model

$$Y = \mu + \sum_t \{\alpha_{m_{(t)}}(t) + \beta_{f_{(t)}}(t)\} + \epsilon, \tag{3.5}$$

where Y is for a phenotype, μ is the population mean; t represents a QTL whose location is unknown in practice; $\alpha_{m_{(t)}}$ is the maternal effect; $\beta_{f_{(t)}}$ is the paternal effect; ϵ is the residual effect. In addition, $m(t)$ is an indicator of whether mother's maternal allele [$m(t) = 1$] or mother's paternal allele [$m(t) = 2$] has been passed to the offsprings. Similarly, $f(t)$ is an indicator of whether father's maternal allele [$f(t) = 1$] or father's paternal allele [$f(t) = 2$] has been passed to the offsprings. It is assumed that, at a given locus t, we have P$\{m(t) = 1\} = $ P$\{m(t) = 2\} = 1/2$ by Mendel's rule of segregation. The same distribution is assumed for $f(t)$ as well, and $m(t)$ and $f(t)$ are independent. Furthermore, by the random-mating assumption, for two loci: $t_1 < t_2$ separated by the genetic distance $\Delta(cM)$, according to Haldane's formula [Haldane (1919)], the recombination rate is computed as $\theta = (1/2) * (1 - e^{-.02\Delta})$, and P$\{m(t_2) = 1|m(t_1) = 2\} = $ P$\{m(t_2) = 2|m(t_1) = 1\} = \theta$, P$\{m(t_2) = 1|m(t_1) = 1\} = $ P$\{m(t_2) = 2|m(t_1) = 2\} = 1 - \theta$. It is also assumed that the maternal and paternal random effects $\alpha_1(t), \alpha_2(t), \beta_1(t), \beta_2(t)$ are independent and normally distributed with mean zero and variance $0.5\sigma_a^2(t)$, where $\sigma_a^2(t) = \text{var}\{\alpha_1(t) + \beta_1(t)\} = 2\text{var}(\alpha_1(t))$ is called the linkage effect. The residual error ϵ is assumed to be normally distributed with mean zero and variance σ_e^2, and is independent with the random effects.

Under the assumed model, the marginal distribution of Y has $E(Y) = \mu$ and $\text{var}(Y) \equiv \sigma_Y^2 = \sigma_e^2 + \sum_t \sigma_a^2(t)$. As mentioned, variance components are often used to model the covariance structure of the phenotypes of members with a pedigree given their IBD sharing matrix. Considering sibships with s siblings (for sib-pair, $s = 2$), let $Y = (Y_1, \cdots, Y_s)'$ be the phenotypes of the s siblings. Also let $\nu_{ij}(t)$ be the IBD sharing number at locus t of the ith and jth siblings in a sibship. Then $\nu_{ii}(t) = 3$ and we have $\nu_{ij}(t) = 0$ if $m_i(t) \neq m_j(t)$, $f_i(t) \neq f_j(t)$; $\nu_{ij}(t) = 1$ if $m_i(t) = m_j(t)$, $f_i(t) \neq f_j(t)$; or $m_i(t) \neq m_j(t)$, $f_i(t) = f_j(t)$; $\nu_{ij}(t) = 2$ if $m_i(t) = m_j(t)$, $f_i(t) = f_j(t)$. Hence $P\{\nu_{ij}(t) = 0\} = 1/4$; $P\{\nu_{ij}(t) = 1\} = 1/2$; $P\{\nu_{ij}(t) = 2\} = 1/4$. Thus, by model (3.5), we have (Exercise 3.6)

$$\text{cov}\left\{Y_i, Y_j \mid [\nu_{ij}(t)]_{t \in QTLs}\right\} = \frac{1}{2}\sum_t \nu_{ij}(t)\sigma_a^2(t) + \gamma_e\sigma_e^2, \quad i \neq j, \qquad (3.6)$$

where γ_e is the residual correlation between two sibs, assuming the correlation is the same for any sib-pairs. Under the normality assumption, we then have

$$Y = \begin{bmatrix} Y_1 \\ \vdots \\ Y_s \end{bmatrix} \Bigg| \; \nu_{ij}(t), 1 \leq i, j \leq s, t \in QTLs \sim N_s\left(\begin{bmatrix} \mu \\ \vdots \\ \mu \end{bmatrix}, \Sigma\right),$$

where $\Sigma(i, i) = \sigma_Y^2 = \sum_t \sigma_a^2(t) + \sigma_e^2$, $i = 1, \cdots, s$; and

$$\Sigma(i, j) = \frac{1}{2}\sum_t \nu_{ij}(t)\sigma_a^2(t) + \gamma_e\sigma_e^2,$$

$1 \leq i \neq j \leq s$. For simplicity, we assume, for now, that the QTLs are on the markers, and the number of siblings in each pedigree is 2 ($s = 2$) and $\mu = 0$. Squared difference and sum transformations are applied to the phenotypes, that is, at a given ith sib-pair, we let

$$z_{i1} = \left(\frac{Y_{i1} - Y_{i2}}{\sqrt{2}}\right)^2, \quad z_{i2} = \left(\frac{Y_{i1} + Y_{i2}}{\sqrt{2}}\right)^2. \qquad (3.7)$$

Note that under the normality assumption, we have $z_{i1} \perp z_{i2}$, i.e., the two random variables are independent (Exercise 3.7). Furthermore, we have

$$E\{z_{i1} \mid \nu_i(t), t \in QTLs\} = \sigma_Y^2 - \gamma_e\sigma_e^2 - \frac{1}{2}\sum_{t \in QTL's} \nu_i(t)\sigma_a^2(t),$$

$$E\{z_{i2} \mid \nu_i(t), t \in QTLs\} = \sigma_Y^2 + \gamma_e\sigma_e^2 + \frac{1}{2}\sum_{t \in QTL's} \nu_i(t)\sigma_a^2(t) \qquad (3.8)$$

(Exercise 3.7). Here the parameters are subject to the constraints $\sigma_a^2(t) \geq 0$, $\sigma_Y^2 \geq 0$, $\sigma_e^2 \geq 0$, and $-1 \leq \gamma_e \leq 1$. Let $z_i = (z_{i1}, z_{i2})'$ represents the vector of the transformed phenotypes for the sib-pair. Then, the above model can be written in the form of regression $z = X\beta + \epsilon$, where z is the vector of transformed phenotypes $(z_1, \ldots, z_n)'$ for n sib-pairs; X is a function of the IBD sharing numbers (i.e., ν_k, the IBD of the kth sib-pair, $k = 1, \cdots, n$), and β is the vector consisting of σ_Y^2, γ_e, σ_e^2, and $\sigma_a^2(t)$, $t \in QTLs$.

Although variance-component linkage analysis (based on the maximum likelihood) is more powerful than the regression-base approach, especially when the normality assumption holds, implementation of the maximum likelihood method involves intensive computation. In fact, even the evaluation of the likelihood function could be very computationally expensive. Here, the calculation of the likelihood function is based on the joint probability of the IBD-sharing status among all of the sib-pairs.

On the other hand, it can be shown (Exercise 3.8) that the transformed phenotypes given the IBD can be expressed as

$$z_{ij} = (x'_{ij}\beta)\xi_{ij}, \quad i = 1, \ldots, n; \quad j = 1, 2, \tag{3.9}$$

where ξ'_{ij}s are i.i.d $\sim \chi^2_{(1)}$. As a result, z_{ij} is a member of the exponential family, namely the Gamma family (Exercise 3.8). In other words, this is a generalize linear model (GLM) instead of a linear regression model as previous thought. By taking advantage of this new finding, our results through a series of simulation studies are quite impressive (see below). More specifically, we consider two cases: (i) QTLs reside on markers; (ii) QTLs in the middle of their flanking markers. Furthermore, it is assumed that the IBD status are assumed known.

3.3.2 *QTLs on markers*

The feasibility of using the linkage approach for the dissection of a complex disease depends on a number of factors. These include the strength of the genetic components; the number of genes involved and the magnitude of their individual effects, implicitly related to the heritability of the trait (the ratio of the variance attributed by genetic effects to the total variance - the variance of the phenotype); the genetic maps used in the study, etc. As a result, mapping of complex diseases usually requires large data sets to provide sufficient power in order to unravel genetic effects.

In our simulation study, we investigate three cases: 2-QTL, 3-QTL, and 4-QTL. In the first two cases, the number of sib pairs, n, is chosen as

5,000. For the last case, $n = 7500$. The whole human genome excluding the sex chromosome, is examined. On each of these twenty-two autosomal chromosome pairs, the genetic distance Δ is set up as 10 cM between any two markers; therefore, the recombination rate is $\theta = .09$. It is assumed that there is no *crossover interference*. Thus, there are 16 equally spaced markers on each chromosome, assuming that each chromosome is 150 cM long. The true parameters are $\mu = 0$, $\sigma_e^2 = 0.4$, $\sigma_y^2 = 1$ and $\gamma_e = 0.1$. To maintain the heritability $h^2 = 60\%$ (Exercise 3.9), the locations and variances of random effects of the QTLs are set up in each case as follows. (i) case I (2-QTL): In this case, the locations of the QTLs are on the 4th and 15th markers of the 1st chromosome, and the variance of random effects for these QTLS are $\sigma_a^2(4) = \sigma_a^2(15) = 0.3$. (ii) case II (3-QTL): In this case, the locations of the first two QTLs are the same as in case I, while the location of the third QTL is on the 4th marker of the 2nd chromosome; and the variances of the random effects for these QTLs are $\sigma_a^2(4) = \sigma_a^2(15) = \sigma_a^2(20) = 0.2$. (iii) case III (4-QTL): In this case, the locations of the first three QTLs are the same as in case II, while the location of the fourth QTL is on the 15th marker of the 2nd chromosome; and the variances of random effects for these QTLs are $\sigma_a^2(4) = \sigma_a^2(15) = \sigma_a^2(20) = \sigma_a^2(31) = 0.15$. The number of bootstrap samples used for the SAF is 100.

Our goal is to identify the QTLs that influence the variation of the trait or, equivalently, to find the subset of markers that are associated with the phenotypic difference. This may be viewed as a model selection problem in a non-conventional situation in that the phenotypes are correlated within the pedigree (sib-pair). Due to the high dimensional and complex structure of the data, to keep the computational cost manageable, a screening step is proposed to remove irrelevant markers before the SAF is applied, as follows: (a) First compute $Q(M)$, where M is a model with a single marker and $Q(M)$ is the negative log-likelihood under the GLM (Gamma family). Consider the top K markers in terms of having the minimum $Q(M)$, where K is chosen based upon an initial guess or *a prior* knowledge, if possible. (b) Fit the GLM with the K markers selected in (a) to obtain estimates of the coefficients corresponding to these markers. The top half of these markers will be selected based upon the ordered magnitude of the estimates.

After the screening step, the SAF is applied to selected the markers. If the number of selected markers by the fence procedure is close to $K/2$, in other words, if the number of selected markers is close to the boundary, it is recommended that one increase K and go back to perform the screening step. On the other hand, if the number of selected markers after the SAF is

Table 3.4 **The Case of QTLs on the Markers.** *Heritability* 60%. *E - marker detected as QTL is a QTL; B.1 - marker detected as QTL is within 10 cM from a QTL; B.2 - marker detected as QTL is within 20 cM from a QTL. In case more than one markers are detected under B.1 or B.2, one is considered as a true QTL, the other one(s) as false positive QTL(s).*

Case	n	Summary	E	B.1	B.2
I	5000	% true model detected	100	100	100
		mean(sd) # true positive	2(.00)	2 (.00)	2 (.00)
		mean(sd) # false positive	0(.00)	0(.00)	0(.00)
II	5000	% true model detected	92	98	98
		mean(sd) # true positive	2.94(.24)	3(.00)	3(.00)
		mean(sd) # false positive	.08(.27)	.02(.14)	.02(.14)
III	7500	% true model detected	80	91	91
		mean(sd) # true positive	3.86(.32)	3.99(.1)	3.99(.1)
		mean(sd) # false positive	.22(.44)	.1(.29)	.1(.29)

far below $K/2$, the procedure ends and one keeps all the identified markers as the optimal model.

Table 3.4 shows perfect results in case I (2-QTL). The results remain very impressive in case II (3-QTL), having perfect detection under the B.1 standard with very small mean false positives. The probability of correct detection in case III is still quite high (almost 90 % under the B.1 standard, and close to 80 % under the E standard). The mean true positives in this case remains almost perfect (3.99; the target is 4) under the B.1 standard; although the mean false positives has increased compared with case II, it is very small compared with the corresponding mean true positives.

3.3.3   *QTLs at middle of flanking markers*

This time, the set-up is similar to the previous one except that the locations of the QTLs are no longer at markers. More specifically, the QTLs now lie in the midpoint of two flanking markers. This means that, in the 2-QTL case, the first QTL is in the middle of the 3rd and 4th markers of the 1st chromosome; similarly, for the second QTL, its flanking markers are the 14th and 15th of the 1st chromosome. In the 3-QTL case, the locations of the first two QTLs are the same as in the 2-QTL case; while third QTL is in the middle of the 3rd and 4th markers of the 2nd chromosome. Finally, in the 4-QTL case, the locations of the first three QTLs are the same as in the 3-QTL case; while the fourth QTL is in the middle of the 14th and 15th markers of the 2nd chromosome.

Once again, the simulation results, presented in Table 3.5, show the outstanding performance of the SAF. Note that in this case, the true model

Table 3.5 **The Case of Flanking Markers.** *Heritability is* 60%. *E - marker detected is either one of the flanking markers of a QTL; B.1 - marker detected is tightly linked to a QTL that is within* 10 *cM from either one of the flanking markers of the QTL; B.2 - marker detected is a marker tightly linked to a QTL that is within* 20 *cM from either one of theflanking markers of the QTL. In case more than two are detected under B.1 or B.2, two of them are considered correct, the other one(s) is/are considered false positives.*

Case	n	Summary	E	B.1	B.2
I	5000	% true model detected	87	95	96
		mean(sd) # true positive	1.98(.14)	1.99 (.1)	1.99 (.1)
		mean(sd) # false positive	.14(.37)	.05(.21)	.05(.21)
II	5000	% true model detected	84	96	98
		mean(sd) # true positive	2.85(.35)	2.97(.17)	2.98(.14)
		mean(sd) # false positive	.17(.4)	.05(.26)	.03(.22)
III	7500	% true model detected	85	91	92
		mean(sd) # true positive	3.89(.31)	3.96(.19)	3.97(.17)
		mean(sd) # false positive	.17(.42)	.1(.33)	.07(.35)

is not among the candidates. Still, the SAF is capable of detecting the best approximating models by choosing the correct flanking markers.

3.4 Exercises

3.1. Consider the Fay-Herriot model introduced in Example 3.1. Show that the maximum likelihood (ML) equation has a closed-form solution for A, if the D_i's are equal, say, $D_1 = \cdots = D_n = D$. However, if the D_i's are not equal, the ML equation may not have a closed-form solution for A.

3.2. Show that, under the Fay-Herriot model of Example 3.1, we have

$$Q(M) - Q(M_f) = \frac{n}{2} \log \left(\frac{|P_{X^\perp} y|^2}{|P_{X_f^\perp} y|^2} \right)$$

$$= \frac{n}{2} \log \left(1 + \frac{K-p}{n-K-1} F \right),$$

where $p+1$ is the number of columns of X, and $F \sim F_{K-p,n-K-1}$.

3.3. Write an algorithm to numerically compute $s(M, M_f)$ for the fence regarding the Fay-Herriot model (Example 3.1).

3.4. Consider the nested-error regression model of Example 3.3. In this exercise, you are asked to compare the finite-sample performance of the SAF (see Section 3.2) with that of the adaptive fence (Section 3.1). Here $Q(M)$ is chosen as the negative log-likelihood (Example 1.3). Recall that, in this case, a reasonable choice of $s(M, M_f)$ is $\sqrt{(|M_f| - |M|)/2}$ (Section 1.4), where M_f is the model with all the candidate auxiliary variables,

$x_{ijk}, k = 0, 1, \ldots, 5$. Under the same simulation setting as Example 3.3, compare the empirical probabilities of correct selection of the SAF, based on (3.2), and the adaptive fence, based on (1.20), with $\tilde{M} = M_f$ and $s(M, M_f)$ given above.

3.5. Consider, once again, Example 3.3. To make it simpler, suppose that the candidate models include at least the intercept but no quadratic terms. Thus, $\mathcal{M} = \{M_0, M_1, M_2, M_3\}$, where $M_0 = \{1\}$ (i.e., the intercept only), $M_1 = \{1, x_1\}$, $M_2 = \{1, x_2\}$, and $M_3 = \{1, x_1, x_2\}$. Suppose that the values of x_1, x_2 are generated from the $N(0, 1)$ distribution, then fixed throughout, as in Example 3.3; and the true model is M_1.

a. Let $\mathcal{M}_c \subset \mathcal{M}$ and $\mathcal{M}_{ic} \subset \mathcal{M}$ denote the subclasses of correct models and incorrect models, respectively. Determine $\mathcal{M}_c$ and $\mathcal{M}_{ic}$.

b. Generate 100 data sets with $m = 15$ under M_1, with $\beta_0 = \beta_1 = \sigma_v^2 = \sigma_e^2 = 1$. Let $d(M) = Q(M) - Q(M_f)$, where $M_f = M_3$. Make a histogram of (1) $d(M), M \in \mathcal{M}$; (2) $d(M), M \in \mathcal{M}_c$; and (3) $d(M), M \in \mathcal{M}_{ic}$, with the same range for the x-axis, one histogram on top of the other. How many modes do you see in histogram (1)? Do you see a separation, in these histograms, between $\mathcal{M}_c$ and $\mathcal{M}_{ic}$? If so, does the separation suggest something on how to choose the cut-off c in (3.2)?

c. For each of the data sets generated in part **b**, obtain the value of the adaptive c in SAF, say, c^*. You may use $B = 100$ as the number of bootstrap samples for the SAF. What is the mean of c^* over the 100 data sets? Comment on the value in view of the histograms in part **b**.

3.6. Verify the covariance identity (3.6).

3.7. Show that the two random variables defined in (3.7) are independent under the normality assumption. Also verify the expressions (3.8).

3.8. Continue with Exercise 3.7.

a. Verify the result (3.9).

b. Show that, conditional on IBD, z_{ij}'s are independent and $\sim \Gamma(1/2, 2a)$ where $a = x'_{ij}\beta$.

c. Show that $\mathrm{E}(z_{ij}|X) = x'_{ij}\beta$, and $\mathrm{var}(z_{ij}|X) = 2(x'_{ij}\beta)^2$, where X is a function in IBD.

d. Thus, according to McCullagh & Nelder (1989, p. 294), z_{ij} follows a GLM with the identity link function.

3.9. This exercise is regarding the genetic quantity heritability. Suppose that you are unfamiliar with the term (otherwise, you have to pretend that you do not know it), search the literature and then write a one-page article about this quantity. Note that there may be more than one way to define heritability; nevertheless, you should look for the one whose value always lies between 0 and 1.

Chapter 4

Restricted Fence

The SAF discussed in Section 3.2 reduces the computational burden of the adaptive fence by avoiding the evaluation of $s(M, \tilde{M})$. On the other hand, even with this simplification, one may still encounter computational difficulties when applying the fence to high dimensional and complex problems. The main difficulty rests in the evaluation of a large number of $\hat{Q}_M$'s, if, for example, the number of candidate variables is fairly large. More specifically, if all-subset selection is considered, and there are k candidate variables, a total of 2^k evaluations of different $Q(M)$'s are required. For example, for $k = 10$, the total number of evaluations is 1,024; for $k = 30$, the total number of evaluations is 1,073,741,824 (over a billion); and the number goes up quickly. Earlier, effort was taken to avoid the all-subset selection by considering a stepwised fence procedure. See Section 1.5. Note that the amount of computation for the adaptive fence, or SAF, is usually much higher than that for the non-adaptive fence due to the need for bootstrapping. In this chapter, we consider a different approach to making the fence method computationally more attractive.

4.1   Restricted fence procedure

Restricted maximum likelihood (or residual maximum likelihood, REML) is a well-known method in linear mixed model analysis [e.g., Jiang (2007)]. The idea is to apply a transformation to the data in order to get rid of the (nuisance) fixed effects. Maximum likelihood is then applied with the transformed data to estimate the dispersion parameters, or variance components, involved in the linear mixed model. The transformation is constructed in such a way so that there is no loss information by using REML in estimating the variance components.

The idea of restricted fence may be viewed as a combination of REML and the SAF. Our focus is selection of the fixed covariates. First, we apply a transformation to the data that is orthogonal to a (large) subset of the candidate variables to make them "disappear". The SAF is then applied to the remaining (small) subset of variables. Below we describe the method in detail through an example.

Broman and Speed (2002) considered a model selection problem in Quantitative Trait Loci (QTL) mapping in a backcross experiment with 9 chromosomes, each with 11 markers. This led to a (conditional) linear regression model with 99 candidate variables, one corresponding to each marker. Following their approach, we consider a linear regression model conditional on the inheritance matrix $X = (x_{ij})_{1 \leq i \leq n, j \in S}$,

$$y_i = \mu + \sum_{j \in S} \beta_j x_{ij} + \epsilon_i, \quad i = 1, \ldots, n,$$

where y_i is the phenotype of the ith subject, $x_{ij} = 1$ or $x_{ij} = 0$ according to whether the ith subject has the homozygote or heterozygote genotype at the jth marker, and S is a subset of regressors corresponding to the markers. The errors, ϵ_i, are assumed to be independent and identically distributed as $N(0, \sigma^2)$. Write $X_1 = (x_{ij})_{1 \leq i \leq n, j \in S_1}$, and $X_2 = (x_{ij})_{1 \leq i \leq n, j \in S_2}$, where S_1 is a subset of S, and $S_2 = S \setminus S_1$. For example, S_1 may correspond to the subset of markers in the first half of the first chromosome, and S_2 the rest of the markers. Then the model can be expressed, in matrix form, as

$$y = X\beta + \epsilon = X_1 \beta^{(1)} + X_2 \beta^{(2)} + \epsilon,$$

where $y = (y_i)_{1 \leq i \leq n}$, $\beta = (\beta_j)_{j \in S}$, $\beta^{(1)} = (\beta_j)_{j \in S_1}$, $\beta^{(2)} = (\beta_j)_{j \in S_2}$, and $\epsilon = (\epsilon_i)_{1 \leq i \leq n} \sim N(0, \sigma^2 I_n)$. Let $p_j = \mathrm{rank}(X_j)$, $j = 1, 2$. Let A be a $n \times (n - p_2)$ matrix such that

$$A'A = I_{n-p_2}, \quad A'X_2 = 0. \tag{4.1}$$

It follows that $AA' = P_{X_2^\perp} = I_n - P_{X_2}$, where $P_{X_2} = X_2(X_2'X_2)^{-1}X_2'$. Then, we have $z = A'y = \tilde{X}_1 \beta_1 + \eta$, where $\tilde{X}_1 = A'X_1$, and $\eta = A'\epsilon \sim N(0, \sigma^2 I_{n-p_2})$.

Note that, by applying the transformation A' to the data, the matrix X_2, which is typically of much higher dimension, has disappeared from the model. Thus, one can apply the SAF to the subset of markers corresponding to X_1 which is usually in much lower dimension. Also note that, although the matrix A is introduced here, its explicit form is not needed for the application of fence method. For example, if $Q(M)$ is chosen as the residual sum of squares (RSS), then it can be shown (Exercise 4.1) that

$$Q(M) = y' P_{X_2^\perp \ominus X_1} y \tag{4.2}$$

with $P_{X_2^\perp \ominus X_1} = P_{X_2^\perp} - P_{X_2^\perp} X_1 (X_1' P_{X_2^\perp} X_1)^{-1} X_1' P_{X_2^\perp}$. Furthermore, for the SAF, one can bootstrap under the full model restricted to S_1 without having to know or estimate β_2. For example, let S_1 denote the set of markers on the first half of the first chromosome and

$$\begin{aligned}
\hat{\beta}_{f,1} &= (\tilde{X}_{f,1}' \tilde{X}_{f,1})^{-1} \tilde{X}_{f,1}' z \\
&= (X_{f,1}' P_{X_2^\perp} X_{f,1})^{-1} X_{f,1}' P_{X_2^\perp} y.
\end{aligned} \tag{4.3}$$

Then, the bootstrap version of $Q(M)$ is given by (Exercise 4.2)

$$Q^*(M) = (X_{f,1} \hat{\beta}_{f,1} + \epsilon^*)' P_{X_2^\perp \ominus X_1} (X_{f,1} \hat{\beta}_{f,1} + \epsilon^*), \tag{4.4}$$

where ϵ^* is the vector of bootstrap errors ϵ. Note that (4.4) is used to select $M = X_1$ (not $X_{f,1}$, which is fixed during the selection), which corresponds to a subset of S_1.

By applying the restricted fence to each half chromosome we are able to identify the marker(s) that are linked to the QTLs. Note that the restricted fence is performed under the transformed data that are different on each half chromosome. In the end, a number of markers, or none, are picked up from each half chromosome. Let $\mathcal{M}_1$ denote the collection of these picked markers.

Finally, another SAF is applied to $\mathcal{M}_1$ the select the final optimal model (i.e., subset of markers).

A numerical algorithm for RF is outlined below:

1. Determine the division of the candidate variables $\mathcal{S} = \{1, \ldots, J\} = \mathcal{S}^{(1)} \cup \cdots \cup \mathcal{S}^{(q)}$, where $1, \ldots, J$ represent the candidate variables $x_j, 1 \leq j \leq J$, and $\mathcal{S}^{(r)}, 1 \leq r \leq q$ are subsets of $\mathcal{S}$.

2. Let $\mathcal{S}_1 = \mathcal{S}^{(1)}$ and $\mathcal{S}_2 = \mathcal{S} \setminus \mathcal{S}_1$. Apply the SAF using the measure of lack-of-fit (4.2) to select the variables among $x_j, j \in \mathcal{S}_1$. Let the subset of selected variables be S_1^*.

3. Repeat 2 with $\mathcal{S}^{(1)}$ replaced by $\mathcal{S}^{(r)}, r = 2, \ldots, q$. When these are done, we have the subsets of selected variables, $S_r^*, r = 1, \ldots, q$.

4. Apply another SAF, if necessary, to $S^* = \cup_{r=1}^q S_r^*$ to select the final optimal subset of variables.

In the following sections, we consider applications of the restricted fence idea to two specific areas, longitudinal studies and backcross experiments.

4.2   Longitudinal studies

Longitudinal data are frequently encountered in clinical trials, medicine, epidemiology, psychology, education and economics, among other fields. The literature in these fields has contributed to the rapid development of statistical methods for longitudinal data analysis [e.g., Diggle *et al.* (2002)]. Longitudinal data provides an opportunity to study the progresses of characteristics of interest over time. In fact, an assessment of the within-subject changes in the response over time can only be achieved with a longitudinal study design. In contrast, the response in a cross-sectional study is measured at a single occasion, therefore, only estimates of the subject-population characteristics can be obtained. In other words, a cross-sectional study design is not capable of capturing information on the response changes during a course of time.

The objectives of longitudinal data analysis are threefolds. These include: (1) to capture the inter-correlation within the subjects; (2) to separate between *cohort* and *age* effects; and (3) to borrow strength across subjects. Such analyses would allow researchers to distinguish changes over time within individuals (aging effect) from differences among subjects at their baseline levels (cohort effect), and make efficient use of the available information (i.e., borrowing strength across subjects). Furthermore, valid inferences can be made more robust to model assumptions, such as in the generalized estimating equation (GEE) approach, described as follows.

4.2.1   *Inference about parameters of main interest*

In most longitudinal studies, the main interests are associated with the so-called mean response. For example, how does the mean response relate to some of the covariates, such as age, sex, body mass index and blood pressure? How does the treatment (e.g., drug) affect the mean response? And how does the mean response change over time? As mentioned earlier, in longitudinal studies, responses collected from the same individual over time are expected to be correlated. As a result, classical statistical analysis such as linear regression may not be appropriate for longitudinal data. For example, by ignoring the response correlation the standard error calculation may be inaccurate, which leads to falsely rejected or accepted null hypotheses; or confidence intervals with incorrect coverage probability or unnecessarily wide. In fact, this has been the main reason that mixed effects models are widely used in longitudinal data analysis. The simplest model assumes that there is a random effect corresponding to each individual. The responses from the same individual are therefore correlated for sharing the same random effects. On the other hand, as mentioned earlier, the responses from different individuals are assumed to be independent. More generally, these models may be expressed as

$$y_i = X_i\beta + Z_i\alpha_i + \epsilon_i, \quad i = 1, \ldots, n \tag{4.5}$$

where n represents the number of individuals (subjects) involved in the study; y_i represents the vector of responses from the ith individual collected over time; and X_i is the matrix of covariates corresponding to the same individual. Note that some of these covariates may be time dependent (such as blood pressure and blood serum measures); others may not be time dependent (such as sex and age, if the duration of the study is relatively short). Furthermore, β is a vector of unknown regression coefficients which are often related to the question of main interest; Z_i is known as a design matrix, α_i is a vector of random effects associated with the ith individual, and ϵ_i is a vector of additional errors. It is assumed that $y_1, \ldots, y_n$ are independent, but the components of y_i are correlated due to the structure of this model. It is also assumed that the random effects and errors have mean zero. Therefore, the mean response is represented by $X_i\beta$. Typically, the mean response consists of two parts. The first part is a linear function of covariates, which is the same as what one has in linear regression; the second part is time-dependent which may involve a function of time and some time-dependent covariates.

Note that (4.5) is a special case of the linear mixed models. If normality is assumed (i.e., the random effects and errors are normal), the model can be fitted by maximum likelihood or REML. See, for example, Jiang (2007) for details. On the other hand, the mixed model approach requires considerable modeling of the covariance structure of the data, and hence may suffer from model misspecification. An alternative approach is the GEE, which is more robust to misspecification of the covariance structure of the data. This was first proposed by Liang and Zeger (1986). Assume $E(y_i) = X_i\beta$ and $Var(y_i) = V_i$, an unknown covariance matrix. Note that the random effects are not explicitly involved; in other words, the model is a *marginal*. If the V_i's were known, the of β would be given by

$$\tilde{\beta} = \left(\sum_{i=1}^{n} X_i'V_i^{-1}X_i\right)^{-1} \sum_{i=1}^{n} X_i'V_i^{-1}y_i \tag{4.6}$$

(Exercise 4.3). However, the V_i's are usually unknown in real life. Therefore, we replace the V_i^{-1} in (4.6) by a known (symmetric) matrix, say,

W_i, called the working inversed covariance matrix. For example, one may use $W_i = I$, the identity matrix. This replacement does not affect the consistency of the estimator [Liang and Zeger (1986)], but may affect the efficiency of the estimator. Nevertheless, the covariance matrix of the estimator (4.6), with V_i^{-1} replaced by W_i, can be estimated by the following "sandwich estimator":

$$\left(\sum_{i=1}^{n} X_i' W_i X_i \right)^{-1} \left(\sum_{i=1}^{n} X_i' W_i \tilde{V}_i W_i X_i \right) \left(\sum_{i=1}^{n} X_i' W_i X_i \right)^{-1}, \qquad (4.7)$$

where $\tilde{V}_i = (y_i - X_i \tilde{\beta})(y_i - X_i \tilde{\beta})'$ (Exercise 4.4). Moreover, by using an iterative procedure developed by Jiang *et al.* (2007) it is possible to obtain an estimator of β that is asymptotically as efficient as the estimator (4.6) as if the V_i's were known.

4.2.2   *Wild bootstrapping*

While there is an extensive literature on modeling the correlation structures, parameter estimation, and inference about the mean response [e.g., Jones (1993); Hand and Crowder (1995); Diggle *et al.* (2002); Jiang (2007)], longitudinal model selection has received much less attention. In particular, there is a lack of theoretical development regarding model selection criteria due to the nonconventional features of longitudinal data [see Jiang *et al.* (2008)]. Although a practioner may employ a number of heuristic selection criteria, such as AIC [Akaike (1973)], BIC [Schwarz (1978)], HQ [Hannan and Quinn (1979)], and CAIC (or consistent AIC; see Bozdogan (1987)), the theoretical bases for these methods have not been justified in the longitudinal setting. In fact, our simulation results (see Subsection 4.2.3) showed that some of these methods may perform poorly in selecting parsimonious models for longitudinal studies.

On the other hand, the fence method, proposed for non-conventional problems, applies naturally to longitudinal model selection problems. Here, our main focus is selection of covariate variables associated with the fixed effects, that is, the columns of X_i. Selection involving the variance-covariance structure of the longitudinal data will be discussed in the next chapter. As the selection of the covariate variables is often among a considerably large number of candidates, the restricted fence is considered. In order to implement the method, one requirement is the ability to bootstrap. This is fairly straightforward in the illustrative example in Section 4.1. However, under the mixed linear model (4.5), the situation is more complicated.

Ideally, the bootstrapping should be done under the full model of (4.5). To do so, one needs to (a) estimate the parameters, which include the fixed effects β and all the variance components associated with the distributions of α_i and ϵ_i; (b) draw samples $\alpha_i^*, \epsilon_i^*, i = 1, \ldots, n$ from the assumed distributions of α_i and ϵ_i, respectively, treating the estimated variance components as the true parameters; and (c) use $y_i^* = X_{\mathrm{f},i}\hat{\beta}_\mathrm{f} + Z_i\alpha_i^* + \epsilon_i^*, i = 1, \ldots, n$ to generate the bootstrap samples, where $X_{\mathrm{f},i}$ is the covariate matrix under the full model, and $\hat{\beta}_\mathrm{f}$ the estimator of β under the full model. We call such a procedure linear mixed model bootstrapping.

However, there are practical reasons that bootstrapping under the full linear mixed model as above may not be robust. For example, the standard procedures of fitting the linear mixed model (4.5), which are maximum likelihood (ML) and REML, involve numerically solving nonlinear maximization problems or equations. Although these procedures are available in standard software packages, such as SAS, S-plus and R, non-convergence, false convergence, and convergence to local maximums often occur in practice. In such cases, the variance components under the full linear mixed model may be poorly estimated, which results in poor bootstrap approximations, as in step (b) above. This is confirmed, for example, in our simulation studies, in which we found that the restricted fence performs significantly better using the wild bootstrapping method, described below, than using the linear mixed model boostrapping.

The *wild bootstrap* was proposed by Liu (1988) following a suggestion of Wu (1986). Also see Beran (1986). Suppose that we are interested in estimating the mean function of the data, which are independent but not identically distributed. Suppose that we apply the classical bootstrap based on i.i.d. samples from the empirical population to estimate the sampling distribution of the estimator. Can the result still be asymptotically correct? Liu (1988) showed that the answer is yes, even though the classical bootstrap does not seem intuitively appropriate here for the simple reason that the original data are not i.i.d. More specifically, it was shown that this "wild bootstrap" not only captures the first-order limit, but also retains the second-order asymptotic properties in the case of the sample mean. This suggests that the wild bootstrap is robust, at least to some extent, against distributional misspecifications.

In our bootstrapping procedure, we first estimate the fixed effects β_1 under the restricted full model. This is naturally done by minimizing the $Q(M)$ that is the RSS. The estimator is given by (4.3) with $X_1 = X_{\mathrm{f},1}$ and $X_2 = X_{\mathrm{f},2}$, and is denoted by $\hat{\beta}_{\mathrm{f},1}$, where $X_{\mathrm{f},j}$ is the full X_j, $j =$

1, 2. Thus, $X_{\mathrm{f},1}$ corresponds to the restricted full model, which has much fewer covariates than the full model. Here, for simplicity, we assume that $X'_{\mathrm{f},1} P_{X^{\perp}_{\mathrm{f},2}} X_{\mathrm{f},1}$ is non-singular. For example, in our simulation study (see the next subsection), the full model of X has 30 fixed covariates, while the full model of X_1 has either 7 or 8 fixed covariates. Next, we write (4.5) as $y = X\beta + \zeta$, where $y = (y_i)_{1 \leq i \leq m}$, $X = (X_i)_{1 \leq i \leq m}$, and ζ represents the rest of the model involving the random effects and errors. We then assume a *working distribution* for the error vector ζ such that, under the working distribution, the components of ζ are independent and distributed as $N(0, \sigma^2)$, where σ^2 is an unknown variance that is estimated by the standard unbiased estimator,

$$\hat{\sigma}^2 = \frac{Q(M_{\mathrm{f}})}{n - p_{\mathrm{f}}} = \frac{y' P_{X^{\perp}_{\mathrm{f},2} \ominus X_{\mathrm{f},1}} y}{n - p_{\mathrm{f}}}, \tag{4.8}$$

where $Q(M)$ is given by (4.2), and $p_{\mathrm{f}} = p_{\mathrm{f},1} + p_{\mathrm{f},2}$ with $p_{\mathrm{f},j} = \mathrm{rank}(X_{\mathrm{f},j})$, $j = 1, 2$ (Exercise 4.5). Given $\hat{\sigma}^2$, we generate ϵ^* by $\epsilon^* = \hat{\sigma}\xi$, where the components of ξ are generated independently from the $N(0, 1)$ distribution. We then use (4.4) to compute $Q^*(M)$, the bootstrap version of $Q(M)$, for the SAF for selecting the covariates for X_1.

In a way, our case is similar to what Liu considered. The underlying model is a linear mixed model, but we are bootstrapping under a regression model that has the same mean function, asymptotically. In other words, the bootstrap draws samples that have the correct mean vector but incorrect covariance matrix. Thus, we refer our bootstrap procedure also as wild bootstrap, following Liu (1988). By using similar arguments as Liu (1988), it can be shown that our wild bootstrap captures the first-order limit, which is what matters for the consistency of model selection.

There is also a similarity between our wild bootstrap procedure and the GEE method, introduced in the previous subsection. In GEE, the means of the responses are correctly specified but the covariance matrices may be misspecified. Nevertheless, the GEE estimator is consistent, even thought it may not be efficient [Liang and Zeger (1986)]. In our wild bootstrap procedure, the bootstrapped $Q(M)$, given by (4.4), depends on $X_{\mathrm{f},1} \hat{\beta}_{\mathrm{f},1} + \epsilon^*$. The first term is correctly specified. This is because the least squares (LS) estimator of $\beta_{\mathrm{f},1}$, which is a special GEE estimator, is consistent (Exercise 4.6). On the other hand, the covariance matrix of ϵ^* may be misspecified, but this does not affect the consistency property of the model selection. Note that only selection of the fixed covariates are considered here. By a very similar argument as that in Jiang *et al.* (2008) or Jiang *et al.* (2009), the consistency property of the restricted fence using the wild bootstrap can be rigorously established. For the most part, the consistency of fence rests on a single requirement, that is, the values of $Q(M)$ are well-separated between the correct and incorrect models. See Chapter 9 for details. It can be shown that the $Q(M)$ given by (4.2) has the latter property. The technical conditions and proof are omitted; see the Supplementary Appendix of Nguyen and Jiang (2012). The results support our observation that, when the parameter estimation is unreliable due to computational instability, the simple and much more stable wild bootstrap may have an advantage.

4.2.3 *Simulation study*

To demonstrate the performance of the restricted fence using the wild bootstrap, we carry out a simulation study under the following linear mixed model for longitudinal data: $y_{ij} = x'_{ij}\beta + v_i + \epsilon_{ij}$, $i = 1, \ldots, n$; $j = 1, \ldots, T$, where i represents the ith subject, and j the jth time point; $v_i \sim N(0, \sigma_v^2)$, $\epsilon_{ij} \sim N(0, \sigma_e^2)$, and v_i's and ϵ_{ij}'s are independent.

The data were simulated to mimic a longitudinal study regarding a bone turnover study. The data for the study were collected over three time points (within 12 month study period). The participants were women ($18 - 40$ years of age) of two groups, vegan or omnivore. The outcome of interest was a marker of bone formation (Osteocalcin), measured over time with respect to dietary groups. The covariates including 30 variables. The variables are listed in Table 4.1.

As the measurements were collected at three time points, we have $T = 3$. We consider three cases: $n = 50$, $n = 100$, and $n = 150$, where n is the number of subjects. In the real dataset, there were 48 participants, 24 of them are vegan and the others are omnivore. Thus, we set up the simulations in a similar way so that half of the subjects are in each dietary group. The continuous covariates are generated under the normal distribution with the mean and standard deviation equal to those obtained from the real dataset with respect to time and group categories, disregarding the missing values.

The true model used for the simulation includes the following variables: time, dietary group, height, N-telopeptide, and crude calcium balance. The true regression coefficients are $\beta = (1, 1, 1, 1, .05, .25, .001)'$, corresponding to the intercept, two time-point indicators, and the rest of the variables. These coefficients are set to be similar to those obtained from the real data under the full model, except for the coefficient of crude calcium balance.

Table 4.1 **Variable Information.**

#	Code	Explanation
1	Const	1 for intercept
2	Tx	vegan/omnivore
3	Time1	time indicator 1 (6 month - baseline)
4	Time2	time indicator 2 (12 month - baseline)
5	Age	age
6	Wtkg	weight
7	Htcm	height
8	Ntx	N-telopeptide
9	TbBMD	total body BMD
10	TbBMC	total body BMC
11	Tbfat	total body fat
12	SBMD	spine BMD
13	SBMC	spine BMC
14	FBMD	femure BMD
15	FBMC	femure BMC
16	Kcal	calories
17	Protein	protein
18	CH	carbohydrate
19	Fiber	fiber
20	Fat	fat
21	Sat	saturate fat
22	Chol	cholesterol
23	VitA	vitamin A
24	VitC	vitamin C
25	Calc	crude calcium balance
26	Iron	iron
27	SOD	super oxide dismutase
28	RNAE	urinary renal net acid excretion
29	Camg	urinary calcium excretion
30	Mgmg	urinary magnesium excretion

The latter variable is known to be associated with the bone metabolism [Anderson and Garner (1995)]. Yet, it is not a significant variable according to the real-data analysis in that its coefficient under the full model was quite small. Thus, in the simulation, we bring this variable into the true model (for practical interest) by increasing the value of its coefficient up to .001. The variances of the subject-specific random effects and errors, σ_v^2 and σ_e^2, are set to be 1, which is close to their estimates from the real data. The number of bootstrap samples for the restricted fence is 100. A total of 100 simulations are run under each sample size.

For the restricted fence, we divide all potential predictors into four groups according to biological considerations. The variable groups are Group A-1: 1—8; Group A-2: 9—15; Group A-3: 16—22; and Group A-4: 23—30, where the variable numbers correspond to those listed in Table 4.1.

The true variables correspond to the numbers 1, 2, 3, 4, 9, 10, and 27 on the list. We then apply the SAF to each group, as described in Section 4.1. The results are reported in Table 4.2. Same-data comparisons are made with four of the traditional information criteria, namely, AIC, BIC, HQ, and CAIC. Due to the high dimensional and complex data structure, the F-B procedure of Broman and Speed (2002) (see Section 1.5), is applied, where the forward selection stops when 50% of the candidate variables are selected; the forward selection is followed by the backward elimination. We then apply AIC, BIC, HQ and CAIC to the sequence of models generated by the F-B procedure, and choose the model with the minimum AIC, BIC, HQ, or CAIC as the optimal model under the respective criterion.

In addition, there have been several shrinkage variable selection methods, following the Lasso [Tibshirani (1996)]. See Chapter 8 for more details. We make the same-data comparisons of our method with two of the most popular shrinkage methods, namely, the adaptive Lasso [Zou (2006)] and the smoothly clipped absolute deviation method [SCAD; Fan and Li (2001); Fan and Lv (2008)]. It has been shown [Zou (2006)] that the Lasso is not consistent for model selection while the adaptive Lasso is. Therefore, our comparison focuses on the latter. It should also be pointed out that there have been recent work on simultaneous selection of fixed and random effects in linear mixed effects models using the shrinkage methods [Bondell *et al.* (2010); Ibrahim *et al.* (2011)]. However, because our focus is selection of the fixed covariates only, it seems more fair to compare with the shrinkage methods that focus on the fixed covariates, namely the adaptive Lasso and SCAD, even though the latter use regression-based measures of lack-of-fit. Interestingly, all the methods being compared, including our method (see the last paragraph before Subsection 3.1), AIC, BIC, HQ and CAIC, use regression-based measures of lack-of-fit.

Table 4.2 summarizes the performance of the restricted fence comparing with those of the (F-B) BIC, CAIC, HQ and AIC procedures as well as the adaptive Lasso (ALASSO) and SCAD. For the latter two methods we use the software packages provided by the authors of those papers, with the regularization parameters chosen by the BIC for both methods. Overall, the restricted fence seems to outperform, significantly, all the other procedures, both in terms of the (empirical) probability of correct selection and in terms of the (empirical) mean and standard deviation of the number of incorrectly selected variables. Some plots (obtained from a randomly selected simulation) of p^* vs. c are shown in Figures 4.1 and 4.2.

Table 4.2 **Empirical Probabilities, Means and Standard Deviations.**
The true model includes 7 variables. *RF - Restricted Fence procedure; BIC -
the F/B BIC procedure, CAIC - the F/B consistent AIC procedure, HQ - the
F/B Hannan & Quinn procedure, AIC - the F/B AIC procedure, ALASSO - the
adaptive LASSO procedure, SCAD - the SCAD procedure. Underfitting (UF)
- the case that at least one true variable is missing in the selected model, but
may include extraneous variable(s). Overfitting (OF) - the case that the selected
model includes all the true variables plus at least one extraneous variable. TP
- detecting (exactly) the true variables. TP, UF and OF in % of empirical
probability. MC - empirical mean # of correct variables in the selected model;
MIC - empirical mean # of incorrect variables in the selected model. Standard
deviations (s.d.) in the parentheses.*

n	Summary	RF	BIC	CAIC	HQ	AIC	ALASSO	SCAD
50	TP	53	37	33	28	0	15	0
	UF	26	46	39	39	14	32	29
	OF	21	17	28	33	86	53	71
	MC	6.73	6.54	6.61	6.61	6.86	6.66	6.63
	s.d.	0.46	0.50	0.49	0.49	0.34	0.52	0.80
	MIC	0.34	0.39	0.68	0.77	4.28	1.84	1.30
	s.d.	0.60	0.69	0.88	0.88	1.93	1.98	0.48
100	TP	85	65	54	47	0	30	0
	UF	5	7	4	3	1	2	44
	OF	10	28	42	50	99	68	56
	MC	6.94	6.92	6.96	6.97	6.99	6.98	6.28
	s.d.	0.27	0.30	0.19	0.17	0.10	0.14	1.05
	MIC	0.12	0.33	0.52	0.72	4.70	1.76	6.72
	s.d.	0.38	0.53	0.65	0.87	2.03	2.05	0.58
150	TP	96	82	69	56	0	37	0
	UF	3	0	0	0	2	1	86
	OF	1	18	31	44	98	62	14
	MC	6.97	7.00	7.00	7.00	6.98	6.99	5.39
	s.d.	0.17	0.00	0.00	0.00	0.14	0.10	1.43
	MIC	0.01	0.22	0.38	0.58	4.14	1.45	10.58
	s.d.	0.10	0.50	0.63	0.76	2.01	1.85	0.93

4.2.4 *Discussion*

The restricted fence begins with the division of the candidate variables into
several groups. We have indicated that the grouping is based on biological
information. Nevertheless, there is concern on sensitivity of the variable
selection result to the grouping. Suppose that the biological information is
ignored in the grouping. Will the result be dramatically different? We carry
out additional simulation studies to investigate this problem. Recall that,
so far in the simulation study, the candidate variables were divided into
four groups, A-1—A-4, according to the biological considerations. Call this
grouping Strategy A. In the additional simulation study, we ignore the bio-

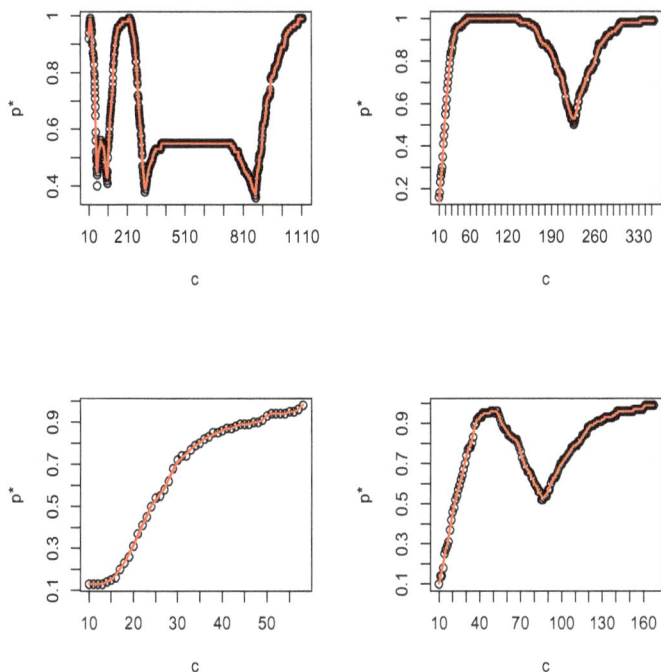

Fig. 4.1 *Plots from step 1 of the restricted fence in one simulation. The four p^* vs. c plots correspond to the four bins of variables. Five variables are picked up from the upper left plot; one variable is selected from the upper right plot; no variable selected from the lower left plot; and one more variable is picked from the lower right plot.*

logical consideration (which is something that we would not recommend in practice). Instead, we consider a different grouping strategy, called grouping strategy B, by shuffling the variable numbers randomly (keeping 3 and 4 together, which are the time-point indicators—it does not make sense to separate them). We then divide the variables into four groups, while maintaining the same numbers of variables in those groups (i.e., 8, 7, 7, 8). The new groups are Group B-1: 5, 8, 16, 18, 22, 23, 24, 28; Group B-2: 1, 2, 9, 10, 19, 20, 21; Group B-3: 3, 4, 13, 14, 17, 29, 30; and Group B-4: 6, 7, 11, 12, 15, 25, 26, 27. We then run the simulations based on the new grouping. The results, corresponding to the middle part of Table 4.2 (i.e., $n = 100$) are TP: 77; UF: 2; OF: 21; MC: 6.76 (.49); MIC: .02 (.14). (The corresponding results for the competing methods do not change, of course.) Comparing with Table 4.2, it is seen that grouping makes some difference,

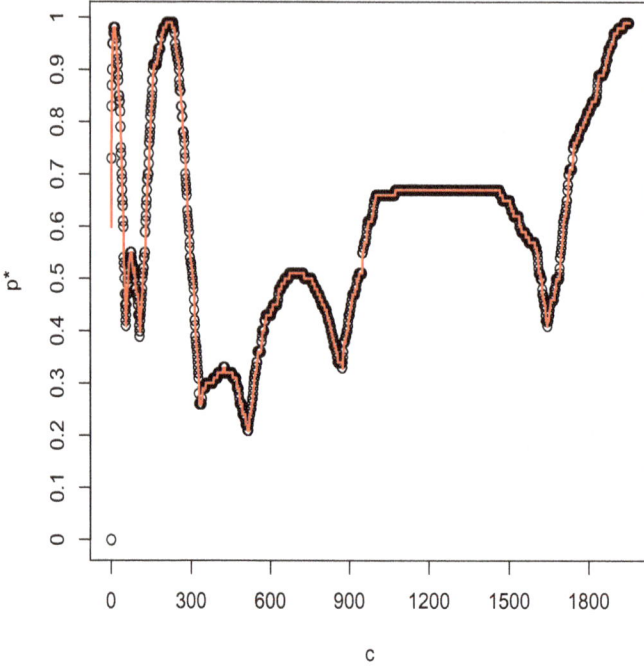

Fig. 4.2 *Plot p* vs. c from step 2 of the restricted fence procedure of one simulation. Seven variables selected from this plot.*

which suggests that information such as biological interest may help. On the other hand, even with the completely randomized grouping, the results have not changed dramatically; in particular, the restricted fence still outperforms the competing methods. This suggests robustness, at least to some extent, for the restricted fence with respect to the grouping.

The robustness of the restricted fence can be argued theoretically in large sample. Because of the consistency of the restricted fence [Jiang *et al.* (2008)], in large sample, the procedure will select the correct variables (and nothing else) with high probability, regardless of the grouping. Equivalently, one may argue in terms of *signal consistency* [Jiang *et al.* (2011b)], which is more appropriate in cases where the number of variables is comparable to the sample size. For the most part, signal consistency means that, as the signals (i.e., the absolute values of the true regression coefficients) increase (but with the sample size fixed), the probability of identifying the true variables (and nothing else) goes to one. As argued in Jiang *et al.*

Table 4.3 **Empirical Results under Stronger Signals.** The true model includes the same 7 variables as in Table 4.2. $n = 100$. Notations are the same as in Table 4.2. The results are exactly the same under the two grouping strategies, A and B.

Summary	RF	BIC	CAIC	HQ	AIC	ALASSO	SCAD
TP	100	70	57	49	0	62	0
UF	0	0	0	0	1	0	100
OF	0	30	43	51	99	38	0
MC	7.00	7.00	7.00	7.00	6.99	7.00	4.00
s.d.	0.00	0.00	0.00	0.00	0.10	0.00	0.00
MIC	0.00	0.33	0.52	0.71	4.70	0.92	6.24
s.d.	0.00	0.53	0.65	0.88	2.03	1.60	0.69

(2011b), it can be shown that the restricted fence is signal-consistent, regardless of the grouping.

To verify the signal consistency of the restricted fence empirically, we consider again the two different grouping strategies. We run simulations with $n = 100$ and the following increased signals for the true variables: 1, 1, 1, 1, .5, .5, and .01 (in other words, the first four coefficients are unchanged, the 5th and 7th are 10 times as strong, and the 6th is twice as strong). The simulation results are presented in Table 4.3. In particular, the results for the restricted fence are exactly the same (which are perfect) under the two grouping strategies, A and B, indicating signal consistency of the restricted fence as aforementioned. Interestingly, the results also seem to suggest that the competing methods improve at a much slower rate as the signals increase, compared to the restricted fence.

It should be noted that consistency or signal consistency are theoretical properties indicating what to expect in the "ideal" situations. In the practical and most likely less ideal situations, a careful design for the grouping could make a difference, as is shown. In short, if there is knowledge about the candidate variables, such as biological interests, the knowledge should be used in the grouping. This was illustrated in Section 4.4 with a real-data example. For the most part, we recommend that the (bio)statistician work closely with the expert scientist(s) in the field to determine what grouping strategy is reasonable. It is also important to take into account the relationships between the variables. Keep in mind that the groups need not be disjoint. Finally, for computational efficiency, the group sizes should be kept relatively small, typically 5—10 variables in each group, if possible.

4.3   Backcross experiments

Unraveling the genetic influences on the phenotypic difference in human is often difficult due to the genetic and cultural heterogeneity of patient populations. One approach to this problem is to use appropriate animal models to pinpoint candidate genes for more focused further investigation. For example, the mouse model has been extensively used in quantitative trait loci (QTL) mappings. A mouse is relatively inexpensive to maintain and breed. The importance of mice in genetic studies was first recognized in the inter-disciplinary field–biological immunology and cancer research [Siegmund and Yakir (2007)]. Today, developed mouse strains serve as primary models for many studies in human diseases, such as obesity, diabetes and cancer. Ideally, the objective of the experiment is to identify susceptible genes that contribute to the variation of a phenotype. However, in practice, such an ambitious aim is often reduced to identifying the genomic regions in which attributed genes are lying. Therefore, it is desirable to form an association between the phenotypic variation and such informative genomic regions.

   Classical methods of QTL mappings includes interval mapping [Lander and Botstein (1989)], composite interval mapping [Zeng (1993, 1994)] and multiple QTL mapping [MQM; e.g., Jansen (1993); Jansen and Stam (1994)]. Recently, model selection methods via the information criteria, e.g., AIC [Akaike (1973)] and BIC [Schwarz (1978)], have received considerable attention in QTL mapping [Broman (1997), Ball (2001), Nakamichi et al. (2001), Piepho and Gauch (2001), Broman and Speed (2002), Silanpää and Corander (2002), Bogdan et al. (2004), Baierl et al. (2006)]. In particular, Broman and Speed (2002) proposed a modification of the BIC criterion, called BIC_δ, to overcome the overestimation of the number of QTLs when using the traditional BIC. However, the modification depends on the size. More specifically, the appropriate value of δ is obtained via extensive Monte Carlo simulations, which changes with the sample size. In other words, unless the sample size of the study coincides with one that has already been considered, the Monte Carlo simulations have to be carried out again in order to determine the value of δ. Bogdan et al. (2004) proposed another version of BIC modification, called mBIC, which includes an extra penalty term that depends on the number of markers and the choice of prior on the QTL numbers. Baierl et al. (2006) modified the mBIC by adjusting its penalty and applied the method to intercross experiments. However all of the BIC modifications depend on several factors, such as the sample size, prior distribution, and the type of experiment; thus, it is not easy to come up with a unified approach along these lines that applies to all the cases.

The fence method, on the other hand, is suitable for non-conventional selection problems. In particular, the adaptive fence is a data driven procedure that does not suffer from the difficulties of the BIC modifications as mentioned above. However, the adaptive fence, or SAF, may encounter computational difficulties when applied to the QTL mapping. This is because the number of markers under consideration is often quite large. For example, Broman and Speed (2002) considered a case involving 99 markers. Thus, the restricted fence is naturally considered to overcome the computational difficulty. In fact, the restricted fence was illustrated in Subsection 4.1 using an example from the backcrossed experiment. To study the problem on a more formal basis, let us first introduce some statistical models for the backcross experiments.

4.3.1 *Statistical models*

Let y_i denote the phenotypic value of individual i, and let $x_{ij} = 1$ (homozygote) or $x_{ij} = 0$ (heterozygote) be the genotype of individual i at marker j. It is assumed that, within the same chromosome, the x_{ij}'s follows a Markov chain with transition probabilities

$$P(x_{i,j+1} = 1 | x_{ij} = 0) = P(x_{i,j+1} = 0 | x_{ij} = 1) = \theta_j,$$

where θ_j is the *recombination fraction* between markers j and $j + 1$; and that $P(x_{ij} = 1) = P(x_{ij} = 0) = 1/2$ in accordance with Mendel's rules. As pointed out by Broman and Speed (2002), identifying the QTLs can be viewed as a model selection problem. First considering the case where QTLs are on the markers, these authors employed an additive model, conditional on the genotypes:

$$y_i = \mu + \sum_{j \in S} \beta_j x_{ij} + \epsilon_i, \tag{4.9}$$

$i = 1, \ldots, n$, where S is a subset of the marker regressors. The errors, ϵ_i, are assumed to be independent and distributed as $N(0, \sigma^2)$. In particular, we consider the above conditional model with 99 candidate variables, one corresponding to each marker.

Similar to Section 3.3, we also investigate the case in which QTLs are no longer at markers. In other words, only markers near but not exactly at functional polymorphisms are genotyped; and even further, QTLs located in the middle of their flanking markers. Consider a simple linear model to test for a QTL located between markers j and $j + 1$,

$$y_i = b_0 + b^* x_i^* + \epsilon_i, \qquad (4.10)$$

$i = 1, \cdots, n$, where b^* is the effect of the putative QTL expressed as a difference in effects between the homozygote and heterozygote, x_i^* is an unobservable indicator variable, taking a value 1 or 0 depending on the genotype of the QTLs under consideration. Given the genotypes of the two flanking markers j and $j + 1$, one has (Exercise 4.7):

$$\begin{aligned}
P(x_i^* = 1 | x_j = 1, x_{j+1} = 1) &= P(x_i^* = 0 | x_j = 0, x_{j+1} = 0) \\
&= (1 - \theta_1)(1 - \theta_2)/(1 - \theta); \\
P(x_i^* = 1 | x_j = 0, x_{j+1} = 1) &= P(x_i^* = 0 | x_j = 1, x_{j+1} = 0) \\
&= \theta_1(1 - \theta_2)/\theta; \\
P(x_i^* = 0 | x_j = 0, x_{j+1} = 1) &= P(x_i^* = 1 | x_j = 1, x_{j+1} = 0) \\
&= (1 - \theta_1)\theta_2/\theta; \\
P(x_i^* = 1 | x_j = 0, x_{j+1} = 0) &= P(x_i^* = 0 | x_j = 1, x_{j+1} = 1) \\
&= \theta_1\theta_2/(1 - \theta), \qquad (4.11)
\end{aligned}$$

where, θ_1, θ_2 are the recombination fractions between markers j, $j + 1$ and the putative QTL, and $\theta = \theta_1 + \theta_2$. In particular, when recombination fraction θ between markers j and $j+1$ is small, we have $\theta \approx \theta_1 + \theta_2$. Since the QTL is in the middle of the two flanking markers, we have $\theta_1 = \theta_2 = \theta/2$.

As noted in Subsection 3.3.3, in the case of QTLs in the middle of flanking markers, there is no true model that is among the candidate models. Thus, in this case, we use the closest approximation (i.e., flanking markers) as the basis for our evaluations of these methods. The performances of the restricted fence, in comparison with the BIC_δ method, are evaluated through simulation studies.

4.3.2   *Simulation studies*

The simulation settings are similar to those in Subsection 3.3.2 and Subsection 3.3.3. The number of progenies is chosen as $n = 500, 750$, or 1000, to study the consistency properties of each procedure. Nine autochromosomes are considered. It is assumed that the length of each chromosome is 100 cM. Eleven equally spaced markers (at a spacing of 10 cM) are genotyped. The recombination process is assumed to be without crossover interference. The marker data are assumed to be complete and without errors. The number of underlying QTLs is either 7 or 9. In 7-QTL case, equal effects, $|\beta| = .76$ at each QTL, are examined. The setting is similar to that of Broman and Speed (2002). The first two QTLs are located at 4th and 8th markers of the 1st chromosome. These two QTLs have *coupling* link (i.e., their effects have the same sign). The next two QTLs are located at 4th and 8th markers of the 2nd chromosome, but have *repulsion* link (i.e., their effects have the opposite sign). The next three QTLs are located at 6th, 4th, and 1st markers of the 3rd, 4th, and 5th chromosomes, respectively. The last four chromosomes contain no QTL. The errors are normally distributed with mean 0 and $\sigma = 1$. As a result, the *heritability* of the trait, defined as the proportion of the variance attributed by genetic effects with respect to the total phenotypic variance is 50% (Exercise 4.8). In the 9-QTL case, the first 7 QTLs are the same as in the previous case, and the last two QTLs are located at the 1st and 6th markers of the 6th chromosome. To maintain the heritability at 50%, the effects of these nine QTLs are reduced to .636. Also, the two new QTLs on the 6th chromosome have a coupling link. There are 100 simulation replicates used in each case.

First consider the case of QTL on the marker. Here the putative markers are completely linked to the QTLs. As mentioned earlier, for complex traits, several susceptible genes may influence phenotypic variation simultaneously, such that the effect of each gene is rather small. As a consequence, it is difficult to detect the QTLs, especially in the case of moderate sample size. Therefore, we propose some practical adjustments while applying the fence methods to genetic applications in general.

First, we smooth the plot of p^* vs. c in order to get rid of small bumps due to the bootstrap sampling, where p^* is the highest empirical probability that a model is selected (see Section 3.2). Here the smoothing function **loess()** in R is used. Figures 4.3–4.5 help to explain why such a smoothing step is helpful. Second, instead of choosing c at which p^* is the highest, we select c based on the "first significant peak" criterion. This adjustment is necessary in the case of weak signals. As discussed in Jiang *et al.* (2008), with an extremely large value of c (which is on the right side of the plot of p^* vs c plot), the model with the lowest dimension is always selected. Now move along the curve from the right to the left, every time some QTLs enter the model one would encounter a peak on the curve. Thus, it is reasonable to think of the leftmost peak as the place where the last QTL(s) is included in the model. Note that, in the case of strong signals, or large sample size, the highest peak is usually also the first peak. This interpretation is supported by Figure 4.5, more specifically, the lower left and right panels.

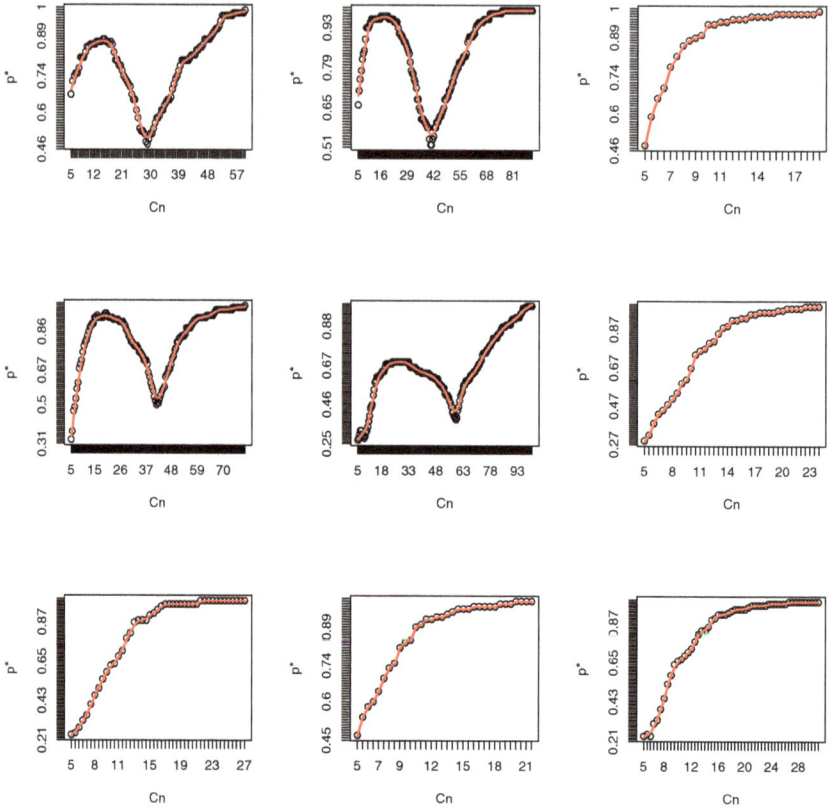

Fig. 4.3   *Plots from step 1 of the restricted fence for the first simulation replicate. Nine p^* vs. $c = c_n$ plots correspond to the nine first halves of the 9 chromosomes.*

The plots for the 9-QTL case are very similar (except that there are more peaks), and therefore omitted.

Results obtained from the restricted fence are compared with those of the BIC_δ method. The authors of the latter method suggested the values of δ as 2.56, 2.1, and 1.85 for $n = 100$, 250, and 500 respectively. However, we obtained the same values (with the same number of simulation runs - 50,000) under the cases of $n = 100$ and $n = 250$ at 97.5^{th} percentile rather than the 95% percentile threshold. Note that, the LOD scores rely on the asymptotic distribution of the likelihood ratio statistic which has an asymptotic χ^2 distribution (not the t-distribution). Therefore, the percentile should be one-sided instead of two-sided. For example, in the case

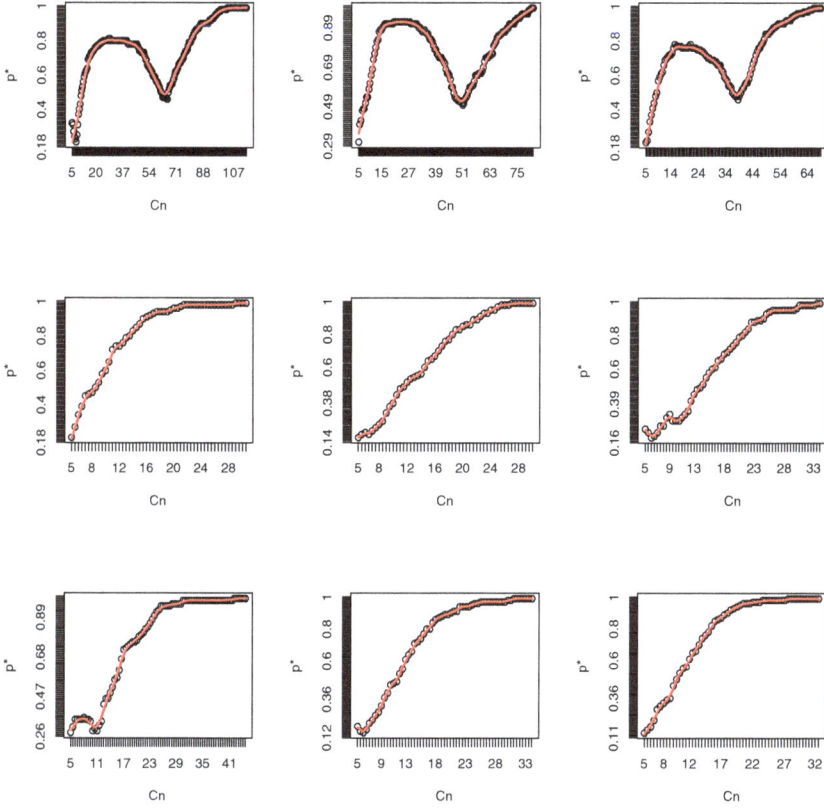

Fig. 4.4 *Plots from step 1 of the restricted fence for the first simulation replicate. Nine p^* vs $c = c_n$ plots correspond to the nine second halves of the 9 chromosomes.*

of $n = 500$, we had $\delta = 1.89$ instead. Note that the result based upon this value is slightly better than that of the choice $\delta = 1.85$. Our values for 95th percentile are 2.23, 1.85, 1.64 for $n = 100$, 250, and 500 respectively. We also obtained δ for the cases of $n = 750$ and $n = 1000$ for our simulation settings. Since the δ values in these two cases are very close to each other, after rounding off (for $n = 750$) and rounding up (for $n = 1000$) with two decimal numbers, they end up with the same values, 1.7 and 1.48 for 97.5^{th} and 95th percentiles, respectively.

Due to the high-dimensionality in QTL mapping, Broman and Speed (2002) incorporated forward selection and backward elimination procedures with BIC_δ. Their most recommended procedure is the forward/backward (F-B) BIC_δ. They also suggested using 25% of the candidate variables

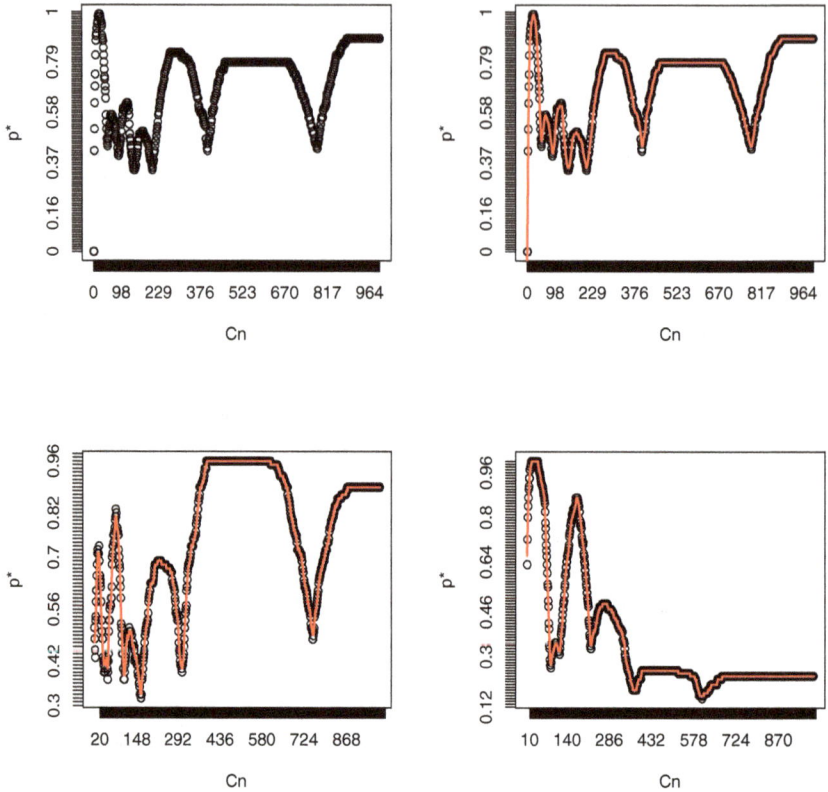

Fig. 4.5   *Plots from step 2 of the restricted fence under the 7-QTL model. Upper left: p^* vs $c = c_n$ of the first simulation replicate with sample size $n = 500$. Upper right: same as upper left with smoothing curve. Lower left: p^* vs $c = c_n$ of one simulation replicate with sample size $n = 500$. Lower right: p^* vs $c = c_n$ of one simulation replicate with sample size of $n = 750$.*

as the stopping rule for the forward selection, and then performing the backward elimination. Thus, in this case, the forward selection continues until 25 markers are picked up, followed by the backward elimination. The optimal model is selected as the one with the minimum BIC_δ among the sequence of models that appear in the F-B procedure. As the signals in our simulations are fairly weak, to avoid missing relevant markers we use instead a 50% stopping rule in the forward selection.

The results for the 7-QTL case are summarized in Table 4.4. In the case of $n = 500$, the restricted fence is very competitive to the F-B BIC_δ. In

Table 4.4 **The 7-QTL Case (QTLs on Markers).** *Heritability* 50%.
*E - marker detected as QTL is a true QTL; B.1 - marker detected as
QTL is within* 10 *cM from a true QTL; B.2 - marker detected as QTL
is within* 20 *cM from a true QTL. In case more than one markers are
detected under B.1 or B.2, one is considered as a QTL; the other one(s)
as false positive QTL(s). RF - Restricted Fence.* $FB.BIC_\delta$ *- F-B* BIC_δ.
$FB.BIC_\delta - 1$: δ *is obtained using the* 95th *percentile of LOD scores.*
$FB.BIC_\delta - 2$: δ *is obtained using the* 97.5^{th} *percentile of LOD scores.*
*%TP - % detecting the true model; M(SD)TP - mean(sd) #'s True pos-
itive; M(SD)FP - mean(sd) #'s False positive. n - sample size.*

Procedure	n	Summary	E	B.1	B.2
$FB.BIC_\delta - 1$	500	%TP	81	90	90
		M(SD)TP	6.89(.34)	7(0)	7(0)
		M(SD)FP	.22(.48)	.11(.34)	.11(.34)
$FB.BIC_\delta - 2$	500	%TP	84	93	93
		M(SD)TP	6.89(.34)	7(0)	7(0)
		M(SD)FP	.19(.46)	.08(.3)	.08(.3)
RF	500	%TP	82	94	94
		M(SD)TP	6.82(.38)	6.94(.23)	6.94(.23)
		M(SD)FP	.16(.36)	.04(.3)	.04(.3)
$FB.BIC_\delta - 1$	750	%TP	89	89	89
		M(SD)TP	7(0)	7(0)	7(0)
		M(SD)FP	.12(.35)	.12(.35)	.12(.35)
$FB.BIC_\delta - 2$	750	%TP	94	94	94
		M(SD)TP	7(0)	7(0)	7(0)
		M(SD)FP	.07(.29)	.07(.29)	.07(.29)
RF	750	%TP	98	100	100
		M(SD)TP	6.98(.38)	7(0)	7(0)
		M(SD)FP	.02(.14)	0(0)	0(0)

fact, the number of true positives for both procedures are almost the same,
yet the number of false positives for the restricted fence is about half of
that for the F-B BIC_δ. For n=750, the fence method is almost perfect in
terms of both true positives and false positives, while the F-B BIC_δ is not
reaching the same state comparatively. The gap of false positives is bigger
in this case, more than 3-fold larger between the F-B BIC_δ and restricted
fence. Overall, the fence method performs better than the BIC_δ. Table 4.5
reports results for the 9-QTL case and the observations are similar.

So far an ideal situation has been considered, that is, the QTLs are
genotyped. However, such a situation is hardly practical. Under a more
realistic scenario, one would genotype markers that are only partially linked
to the QTLs. In other words, only genetic markers near functional polymorphisms are genotyped. Note that the markers do not by themselves
associate with the phenotypic variation. However, since the markers are
linked to the functional polymorphism, which in turn is correlated with

Table 4.5 **The 9-QTL Case (QTLs on Markers).** *Heritability* 50%. *Notation as in Table 4.4.*

Procedure	n	Summary	E	B.1	B.2
FB.BIC$_\delta$ − 1	750	%TP	83	90	90
		M(SD)TP	8.93(0.25)	9(0)	9(0)
		M(SD)FP	.18(.41)	.11(.34)	.11(.34)
FB.BIC$_\delta$ − 2	750	%TP	87	94	94
		M(SD)TP	8.93(.25)	9(0)	9(0)
		M(SD)FP	.14(.37)	.07(.29)	.07(.29)
RF	750	%TP	85	93	94
		M(SD)TP	8.85(.35)	8.93(.25)	8.94(.23)
		M(SD)FP	.11(.31)	.03(.17)	.03(.14)
FB.BIC$_\delta$ − 1	1000	%TP	95	95	95
		M(SD)TP	9(0)	9(0)	9(0)
		M(SD)FP	.05(.21)	.05(.21)	.05(.21)
FB.BIC$_\delta$ − 2	1000	%TP	98	98	98
		M(SD)TP	9(0)	9(0)	9(0)
		M(SD)FP	.02(.14)	.02(.14)	.02(.14)
RF	1000	%TP	99	100	100
		M(SD)TP	8.99(.1)	9(0)	9(0)
		M(SD)FP	.01(.1)	0(0)	0(0)

the phenotype, one may observe associations between the markers and the phenotype. Clearly, the chance of establishing an association between the markers and the trait depends on the strength of the linkage between the functional polymorphism and the markers. Thus, the most challenging situation would be when the QTLs reside in the middle of two markers.

The simulation settings are similar to those of the previous subsection except that the locations of QTLs are no longer at the markers. More specifically, each QTL lies on the midpoint of its two flanking markers. So, in the 7-QTL case, the first QTL is at the midpoint of the 4th and 5th markers of the 1st chromosome. Similarly, for the second QTL, its flanking markers are the 8th and 9th markers of the 1st chromosome, and so on. The 9-QTL case is similar. With the same effects (i.e., coefficients) as in the case that QTL are on the marker, the empirical heritability is now 40% in both the 7-QTL and 9-QTL settings.

The results are summarized in Table 4.6 and Table 4.7. More specifically, as can be seen, the restricted fence performs similarly as BIC_δ in the case $n = 750$, and better in the case $n = 1000$. An important observation is that, unlike the previous case where the QTLs are on the markers, here the performance of BIC_δ does not seem to improve (actually, in most cases, it gets worse) when n increases. In other words, the BIC_δ procedure appears to be inconsistent as n goes to infinity. The authors have mentioned in their

Table 4.6 **The 7-QTL Case (QTLs in Middle of Flanking Markers).** *Heritability* 40%. *E - marker detected is either one of the flanking markers of a true QTL; B.1 - marker detected is within* 10 *cM from either one of the flanking markers of a true QTL; B.2 - marker detected is within* 20 *cM from either one of the flanking markers of a true QTL. In case that more than two markers are detected under B.1 or B.2, two of them are considered as true positives; the other one(s) as false positives. Other notations as in Table 4.4.*

Procedure	n	Summary	E	B.1	B.2
FB.BIC$_\delta$ − 1	500	%TP	54	79	88
		M(SD)TP	6.66(.58)	6.98(.14)	7(0)
		M(SD)FP	.7(.91)	.29(.55)	.23(.52)
FB.BIC$_\delta$ − 2	500	%TP	59	87	93
		M(SD)TP	6.63(.61)	6.97(.17)	7(0)
		M(SD)FP	.57(.79)	.23(.48)	.2(.47)
RF	500	%TP	53	78	85
		M(SD)TP	6.4(.7)	6.77(.5)	6.85(.43)
		M(SD)FP	.56(.8)	.19(.5)	.11(.44)
FB.BIC$_\delta$ − 1	750	%TP	44	76	84
		M(SD)TP	6.93(.25)	7(0)	7(0)
		M(SD)FP	.77(.81)	.43(.63)	.37(.56)
FB.BIC$_\delta$ − 2	750	%TP	53	80	86
		M(SD)TP	6.91(.32)	7(0)	7(0)
		M(SD)FP	.6(.73)	.33(.56)	.79(.51)
RF	750	%TP	76	95	97
		M(SD)TP	6.75(.5)	6.97(.17)	6.99(.1)
		M(SD)FP	.27(.5)	.05(.21)	.03(.17)

paper that the consistency of the procedure has not been studied [Broman and Speed (2002)]. Here consistency means that, as $n \to \infty$, the probability of selecting the optimal model (e.g., exactly the set of true QTLs) goes to one. On the other hand, the performance of the restricted fence improves as n increases, which is consistent with its consistency property that has been established [Jiang *et al.* (2008, 2009), see Chapter 9]. It is remarkable that the restricted fence can select the most parsimonious model even when the true mode is not a member of the candidate models, as in this case.

In addition to the restricted fence and BIC$_\delta$, we also studied performance of the shrinkage variable selection methods, including the Lasso [Tibshirani (1996)], SCAD [Fan and Li (2001)], and adaptive Lasso [Zou (2006)], which are often used in high-dimensional problems. We found that these methods tend to provide much higher false positives rate when applied to the backcross experiments, compared to the restricted fence and BIC$_\delta$. Therefore, the results are omitted [see Nguyen *et al.* (2013)].

Table 4.7 **The 9-QTL Case (QTLs in Middle of Flanking Markers).** *Heritability* 40%. *Notations as in Table 4.6.*

Procedure	n	Summary	E	B.1	B.2
$FB.BIC_\delta - 1$	750	%TP	52	74	82
		M(SD)TP	8.75(.5)	8.96(.19)	9(0)
		M(SD)FP	.7(.89)	.34(.6)	.26(.48)
$FB.BIC_\delta - 2$	750	%TP	60	83	90
		M(SD)TP	8.73(.52)	8.96(.19)	8.99(.1)
		M(SD)FP	.53(.75)	.2(.42)	.13(.33)
RF	750	%TP	58	84	88
		M(SD)TP	8.47(.71)	8.84(.36)	8.88(.32)
		M(SD)FP	.47(.68)	.1(.3)	.06(.23)
$FB.BIC_\delta - 1$	1000	%TP	46	75	89
		M(SD)TP	8.9(.38)	9(0)	9(0)
		M(SD)FP	.75(.9)	.4(.63)	.28(.51)
$FB.BIC_\delta - 2$	1000	%TP	59	87	93
		M(SD)TP	8.9(.38)	9(0)	9(0)
		M(SD)FP	.53(.78)	.26(.5)	.21(.43)
RF	1000	%TP	75	94	95
		M(SD)TP	8.7(.57)	8.95(.21)	8.97(.17)
		M(SD)FP	.3(.65)	.05(.32)	.03(.22)

4.4 A real data example

We illustrate the restricted fence method with a real data example. A clinical trial, Soy Isoflavones for Reducing Bone Loss (SIRBL), was conducted at multi-centers (Iowa State University, and University of California at Davis - UCD). Only part of the data collected at UCD will be analyzed here. The data includes 56 healthly postmenopausal women (45 - 65 years of age) as part of a randomized, double-blind , and placebo-controlled study. The data were collected over three time points–baseline, after 6 and 12 months. One problem of interest is to model the Cytokines (IL1BBLLA, TNFABLLA, IL6BLLA) - inflammatory markers - over time on gene expression for IFNb and cFos, along with other variables listed in Table 4.8. We are interested in finding a subset of relevant variables/covariates that contribute to the variation of Cytokines.

Here we only report the results of data analysis for IL1BBLLA. The covariate variables are grouped into four groups according to biological interest. More specifically, one of the authors worked closely with an expert scientist in the USDA Western Human Nutrition Research Center located at UCD, to determine what variables should be grouped together, and finally came up with the grouping [see Nguyen and Jiang (2012), Supplementary Appendix, for details]. The restricted fence method is then applied in very much the same way as in Subsection 4.2.3. The results are compared with other procedures, reported in Table 4.8.

The main objective of the study was to examine whether Soy Isoflavones treatment affects the bone metabolism. This treatment effect is selected by the restricted fence, AIC and SCAD, but not by the other methods. The Weight variable was thought to be relevant, and is picked up by AIC and HQ, but not by other procedures; however, the BMI variable, which is a function of weight and height, is picked up by the restricted fence and SCAD. As also seen in the same table, BMD for lumbar and spine measures (LSTBMD) is picked up by the restricted fence, but not by any other procedure. Apparently in this analysis, BIC, CAIC, HQ and the adaptive Lasso have over-penalized; as a result, their optimal models do not pick up some relevant covariates, such as BMD and BMC (adaptive Lasso did not pick up any of the variables). As for AIC, it is able to pick up femoral neck area (FNArea) and lumbar spine total area (LSTArea), which are related to bone areal size (i.e., prefix-Area) and considered relevant. However, after consulting with the expert scientist in this field, we are confirmed that BMD and BMC are more important variables than Area measures in this case. Thus, the results of the restricted fence data analysis are more clinically relevant. Although SCAD has selected the most variables, it has missed the important variable LSTBMD. As for the total body area (WBodArea) that is uniquelly picked up by SCAD, the variable is relatively less important, compared to the BMD and BMC, as noted. Our simulation study (see Table 4.2 and Table 4.3) has suggested that SCAD has the tendency of missing important variables as well as selecting extraneous variables.

4.5 Exercises

4.1. Verify (4.2). Note that the expression does not specifically involve A, so long as (4.1) is satisfied.

4.2. Show that the bootstrapped version of $Q(M)$ is given by (4.4), where $\hat{\beta}_1$ is given by (4.3), and ϵ^* is the bootstrapped error.

4.3. Show that, if $E(y_i) = X_i\beta$, $Var(y_i) = V_i$, and the y_i's are independent, the BLUE of β is given by (4.6). Here the BLUE, $\tilde{\beta}$, is in the sense that (i) it is linear in $y = (y_i)_{1 \leq i \leq n}$, that is, $\tilde{\beta} = By$ for some known constant matrix B; (ii) it is unbiased, that is, $E(\tilde{\beta}) = \beta$; and (ii) its covariance matrix is the smallest among all linear unbiased estimators in the sense that $Var(\tilde{\beta}) \leq Var(\hat{\beta})$ for any estimator $\hat{\beta}$ that satisfies (i) and (ii) (with

Table 4.8 **Modeling IL1BBLLA.** *RF - Restricted Fence procedure, BIC - the F/B BIC procedure, CAIC - the F/B consistent AIC procedure, HQ - the F/B Hannan & Quinn procedure, AIC - the F/B AIC procedure, LAS - the Lasso procedure, ALAS - the adaptive Lasso procedure, SCAD - the SCAD procedure. The × indicates variable selected. Variables not listed were not selected by any of the methods.*

Variable	RF	BIC	CAIC	HQ	AIC	LAS	ALAS	SCAD
Soy Treatment	×				×			×
Weight				×	×			
BMI	×							×
WaistCir	×				×	×		×
HipBMD								×
LSTBMC	×					×		×
LSTBMD	×							
TibTrBMC						×		×
TibTrBMD	×	×	×	×	×	×		×
FNArea					×			
LSTArea					×			
WBodArea								×

$\tilde{\beta}$ replaced by $\hat{\beta}$), where for symmetric matrices A, B, $A \leq B$ iff $B - A$ is nonnegative definite.

4.4. This exercise is regarding the derivation of the sandwich estimator (4.7) of $\text{Var}(\tilde{\beta})$, where $\tilde{\beta}$ is given by (4.6) with V_i^{-1} replaced by W_i.

a. Show that the covariance matrix of the estimator (4.6), with V_i^{-1} replaced by W_i is equal to

$$\text{Var}(\tilde{\beta}) = \left(\sum_{i=1}^{n} X_i' W_i X_i \right)^{-1} \left(\sum_{i=1}^{n} X_i' W_i V_i W_i X_i \right) \left(\sum_{i=1}^{n} X_i' W_i X_i \right)^{-1},$$

where $V_i = \text{Var}(y_i)$.

b. Show that

$$\sum_{i=1}^{n} X_i' W_i V_i W_i X_i = \text{E} \left\{ \sum_{i=1}^{n} X_i' W_i (y_i - X_i \beta)(y_i - X_i \beta)' W_i X_i \right\},$$

where β is the true parameter vector so that $\text{E}(y_i) = X_i \beta, 1 \leq i \leq n$.

c. Argue that $\sum_{i=1}^{n} X_i' W_i (y_i - X_i \beta)(y_i - X_i \beta)' W_i X_i = O_P(n)$, while

$$\text{E} \left\{ \sum_{i=1}^{n} X_i' W_i (y_i - X_i \beta)(y_i - X_i \beta)' W_i X_i \right\}$$

$$= \sum_{i=1}^{n} X_i' W_i (y_i - X_i \beta)(y_i - X_i \beta)' W_i X_i + o_P(n),$$

where a sequence of random matrices, ξ_n, is $O_P(n)$ if $n^{-1}\xi_n$ converges in probability to a nonzero limit; and is $o_P(n)$ if $n^{-1}\xi_n$ converges in probability to 0 (matrix) [the actual definition of $O_P(n)$ is somewhat more general; see, for example, Jiang (2010), sec. 3.4]. You may assume some regularity conditions for the argument to hold.

d. Argue that, furthermore, we have

$$\sum_{i=1}^{n} X_i' W_i (y_i - X_i \beta)(y_i - X_i \beta)' W_i X_i$$

$$= \sum_{i=1}^{n} X_i' W_i (y_i - X_i \tilde{\beta})(y_i - X_i \tilde{\beta})' W_i X_i + o_P(n).$$

Again, you may assume some regularity conditions for the argument.

4.5. Show that (4.8) is an unbiased estimator of σ^2 under the working distribution.

4.6. Show that the LS estimator, $\hat{\beta}_{f,1}$, is a special case of the GEE estimator, and it is a consistent estimator of $\beta_{f,1}$.

4.7. Verify the conditional probabilities in (4.11), where $\theta = \theta_1 + \theta_2$.

4.8. Show that, under the simulation settings of both the 7-QTL and 9-QTL cases, described at the beginning of Subsection 4.3.2, the heritability is about 50%.

Chapter 5

Invisible Fence

Another variation of the fence that is intended for high-dimensional selection problems is *invisible fence* [IF; Jiang *et al.* (2011b)]. A critical assumption in Jiang *et al.* (2008) is that there exists a correct model among the candidate models. Although the assumption is necessary in establishing consistency of the fence, it limits the scope of applications because, in practice, a correct model simply may not exist, or exist but not among the candidate models. In this regard, George Box once famously said that "all models are wrong ... but some are useful". Along the same line, we can extend the fence to make it applicable to (much) more general situations. The extension has been done in Section 1.3, where the notions of measure of lack-of-fit and true model were extended. To introduce the IF, we first take another look at the fence methods.

5.1 Another look at the fence

To be specific, assume that the minimum-dimension criterion is used to select the optimal model within the fence. In case there are ties (i.e., two models within the fence, both with the minimum dimension), the model with the minimum dimension and minimum $Q(M)$ will be chosen. Also recall the cut-off c in (3.2). As it turns out, whatever one does in choosing the cut-off (adaptively or otherwise), only a fixed small subset of models have nonzero chance to be selected; in other words, the majority of the candidate models do not even have a chance. We illustrate with an example.

Example 5.1: Suppose that the maximum dimension of the candidate models is 3. Let $M_j^\dagger$ be the model with dimension j such that $c_j = Q(M_j^\dagger)$ minimizes $Q(M)$ among all models with dimension j, $j = 0, 1, 2, 3$. Note that $c_3 \leq c_2 \leq c_1 \leq c_0$; assume no equality holds for simplicity. The point is that any $c \geq c_0$ does not make a difference in terms of the final model selected by the fence, which is $M_0^\dagger$. Similarly, any $c_1 \leq c < c_0$ will lead to the selection of $M_1^\dagger$; any $c_2 \leq c < c_1$ leads to the selection of $M_2^\dagger$; any $c_3 \leq c < c_2$ leads to the selection of $M_3^\dagger$; and any $c < c_3$ will lead to non-selection, because no model is in the fence. In conclusion, any fence methods, adaptive or otherwise, will eventually select a model from one of the four: $M_j^\dagger$, $j = 0, 1, 2, 3$.

The example shows that one may simplify the problem of choosing the optimal model to that of choosing the optimal dimension. Note that there may be thousands, millions, or even billions of candidate models but, in terms of the dimensions of the models, the choices are much more limited. So, at least conceptually, the complexity of the selection is reduced.

It remains to solve the dimension problem. For this we use the AF idea by drawing bootstrap samples, say, under the full model. The idea is to select the model that has the highest empirical probability to best fits the data when controlling the dimension of the model. More specifically, for each bootstrap sample, we find the best-fitting model at each dimension, that is, $M_j^{*\dagger}$, such that $Q^*(M_j^{*\dagger})$ minimizes $Q^*(M)$ all models with dimension j, where Q^* represents Q computed under the bootstrap sample. We then compute the relative frequency, among the bootstrap samples, for different models selected, and the maximum relative frequency, say p_j^*, at each dimension j. Let $M_{j*}^{*\dagger}$ be the model that corresponds to the maximum p_j^* (over different j's) and this is the model we select. In other words, if at a certain dimension we find a model that has the highest empirical probability to best fit the data, this is the model we select. As in Jiang et al. (2008), some extreme cases (in which the relative frequencies always equal to one) need to be handled differently. Although the new procedure might look quite different from the fence, it actually uses implicitly the principle of the AF as explained above. For such a reason, the procedure is called invisible fence, or IF.

5.2   Fast algorithm

Computation is a major concern in high dimension problems. For example, for the 522 gene pathways developed by Subramanian et al. (2005), at dimension $k = 2$ there are $135,981$ different $Q(M)$'s to be evaluated; at dimension $k = 3$ there are $23,570,040$ different $Q(M)$'s to be evaluated, ... If one has to consider all possible k's, the total number of evaluations is

2^{522}, an astronomical number. Jiang *et al.* (2011b) proposed the following strategy, called *fast algorithm*, to meet the computational challenge. Consider the situation where there are a (large) number of candidate elements (e.g., gene-sets, variables), denoted by $1, \ldots, m$, such that each candidate model corresponds to a subset of the candidate elements. A measure Q is said to be *subtractive* if it can be expressed as

$$Q(M) = s - \sum_{j \in M} s_j, \tag{5.1}$$

where s_j, $j = 1, \ldots, m$ are some quantities computed from the data, M is a subset of $1, \ldots, m$, and s is some quantity computed from the data that does not depend on M. Typically we have $s = \sum_{j=1}^{m} s_j$, but the definition does not impose such a restriction. For example, in gene set analysis [GSA; Efron and Tibshirani (2007)], s_j corresponds to the gene-set score for the jth gene-set. As another example, Mou (2012) considered $s_j = |\hat{\beta}_j|$, where $\hat{\beta}_j$ is the estimate of the coefficient for the jth candidate variable under the full model, in selecting the covariates in longitudinal studies.

For a subtractive measure, the models that minimize $Q(M)$ at different dimensions are found almost immediately. Let $r_1, r_2, \ldots, r_m$ be the ranking of the candidate elements in terms of decreasing s_j. Then, the model that minimizes $Q(M)$ at dimension one is r_1; the model that minimizes $Q(M)$ at dimension two is $\{r_1, r_2\}$; the model that minimizes $Q(M)$ at dimension three is $\{r_1, r_2, r_3\}$, and so on (Exercise 5.1).

Below we consider two applications of the IF, implemented via the fast algorithm.

5.3   Gene set analysis

There have been studies on the problem of identifying differentially expressed (d.e.) groups of genes, which we call gene sets, from a set of microarray experiments [e.g., Subramanian *et al.* (2005)]. In particular, Efron and Tibshirani (2007) proposed a gene set analysis (GSA) method based on testing the significance of gene sets. Suppose that there are N genes measured on n microarrays under two different experimental conditions, called control and treatment. The number N is usually large, say, at least a few thousands, while n is much smaller, say, a hundred or fewer. Here the interest is to assess the significance of predefined gene-sets, rather than individual genes, in terms of response to the treatment. The gene-sets are derived from different sources such as biological pathways. See Efron and Tibshirani (2007) for a nice discussion on the gene-set experiments as well as existing methods of gene set analysis. The general procedure of GSA is as follows. First compute a summary statistic for each gene, for example, the two-sample t-statistic. Then, a gene-set statistic is computed for each gene set based on the summary statistics, for example, the average of the summary statistics for genes in the gene-set. The next step is called *restandardization*, by subtracting the genewise mean of the summary statistics from each gene-set statistic and then dividing the difference by the genewise standard deviation of the summary statistics. Results of restandardization are called gene-set scores. The p-values for each gene-set score and false discovery rates (FDR) applied to these p-values are then estimated by permutations of the class labels in the control and treatment. Depending on the FDR cut-off level, a number of the gene sets may be declared significant, or no gene set is declared significant.

An alternative approach to the identification of the d.e. gene sets is to view it as a problem of model selection. Namely, let the candidate gene sets be numbered as $1, \ldots, K$. Then, a collection of the gene sets corresponds to a subset of $\{1, \ldots, K\}$, which we call a model. The goal is to identify, or select, the optimal model, which corresponds to the gene sets that are considered d.e. Note that the problem may be little different from a classical model selection problem in that there may not be a "true model" under this setting. This is because, practically, whether or not a gene set is considered d.e. depends on the available data. The more data (information) one has the more gene sets may be declared d.e. The question is: What is the best subset of d.e. gene sets that the data is able to declare? Nevertheless, the extension of the fence in Section 1.3 has allowed us to apply the IF to this case. Furthermore, a natural subtractive measure is available. Let s_j be the gene-set score corresponding to the jth gene set. For example, s_j may be the maxmean statistic introduced by Efron and Tibshirani (2007). To compute the maxmean, first obtains the positive and negative parts of each genewise summary statistic. Here the positive part of $x \in R$ is $x^+ = \max(x, 0)$, while the negative part is $x^- = -\min(x, 0)$. The means of the positive and negative parts are then taken for each gene set, say, $\bar{s}_j^+$ and $\bar{s}_j^-$, respectively, and the maxmean for the gene set is subsequently defined as $s_j = \max(\bar{s}_j^+, \bar{s}_j^-)$. In their paper, Efron and Tibshirani showed that the maxmeans have the best overall performance as compared to other choices of gene-set scores (such as the means and absolute means). Efron and Tibshirani (2007) suggested using the restandardized maxmeans. Our simulation results (see below), however, reveal some potential problems with the restandardization. The following example suggests another kind of problem, that is, in case of small m, the maxmeans without restandardization are prefered over the restandardized maxmeans.

Example 5.2. (Restandardization when the number of gene-sets is small) Suppose there are m gene sets, each with a single gene, so that the last gene-set is d.e. and the rest are not. For simplicity, suppose that one actually observes the true means of the gene-set scores, which are $0, \ldots, 0, a$, where $a > 0$. Intuitively, one would expect an easier time to detect the d.e. gene-set when a gets larger. If one considers the gene-set scores without restandardization, the difference between the d.e. gene-set and any non-d.e. one is a, which increases with a and is not dependent on m (however, see Exercise 5.2). On the other hand, the restandardized gene-set scores are $-1/\sqrt{m}, \ldots, -1/\sqrt{m}, (m-1)/\sqrt{m}$. So, by restandardization a has disappeared from all of the gene-set scores. In particular, the difference between the d.e. gene set and any of the non-d.e. ones is $\sqrt{m}$, which does not depend on a. However, the difference increases with m, the total number of gene sets. In conclusion, when the number of gene sets is large, restandardization is likely to improve the performance of gene set detection; otherwise, if number of gene-sets is small, restandardization may not be a good idea compared to using gene-set scores without restandardization.

To apply the IF, A subtractive measure for the gene set analysis is given by (5.1) with s_j being the gene-set score. It is clear that all the requirements are satisfied with $s = \sum_{j=1}^{m} s_j$. Therefore, the fast algorithm applies, which simplifies the problem to the determination of the optimal dimension, k, corresponding to the mumber of gene sets that a model involves. As discussed in Section 5.1, the optimal k is determined via a bootstrap procedure similar to the AF, but we need to know how to bootstrap. As in Efron and Tibshirani (2007), the data matrix, X, is $N \times n$, where N is the number of genes and n the number of microarrays, or samples. The underlying assumption is that the samples from the control group are i.i.d., and so are the samples from the treatment group (but the distributions of the samples from the two groups, of course, may be different). Note that each sample is a $N \times 1$ vector of genewise summary statistics. Therefore, a natural approach is to bootstrap separately from the control and treatment groups using the original idea of Efron (1979). For example, suppose there are 50 samples in the control and treatment groups, respectively. We draw a random sample of size 50 with replacement from $1, \ldots, 50$, and a random sample of size 50 with replacement from $51, \ldots, 100$, and then combine the two samples. The columns of X corresponding to the combined sample constitutes the bootstrap sample. This might sound straightforward; however, there is a complication.

As is usually the case in microarray analysis, the dimension N of each sample is much higher than the sample size n. If one has to bootstrap such high dimensional vectors, the quality of bootstrap (in terms of convergence of the bootstrap distribution to the true underlying distribution; see Chapter 9) drops. In other words, bootstrapping the full N-dimensional columns of X may not yield a good approximation to the empirical probabilities used for the determination of k. To solve this problem we use the following strategy called the *limited bootstrap*. The idea is to bootstrap a small number of gene sets initially, and gradually increase the number of bootstrapped gene sets if necessary. Here by bootstrapping the gene sets it means that only the rows of X corresponding to the selected gene-sets (and the columns corresponding to the bootstrap sample) are to be used in determine k. To do this we need to know (i) what gene sets? (ii) how small? and (iii) when to increase?

For (i) we use the following principle: Selection is always made among the top gene sets. Therefore, for any given number, say, l, the gene sets to be bootstrapped are the top l gene sets by the ranking of the gene-set scores.

The answer to (iii) is suggested by the relative frequencies of IF, which are similar to those of the adaptive fence (see Chapter 3). If the highest relative frequency excluding dimension zero occurs at the highest dimension being considered, it is an indication that the number of bootstrapped gene-sets needs to increase.

The answer to (ii) is, again, motivated by the relative frequencies of IF, as follows. Suppose that one needs to carry out an all-subset selection up to dimension K, in which there is 1 model at dimension zero, K at dimension one, $K(K-1)/2$ at dimension two, and so on. Note that the empirical probability or relative frequency at dimension zero is always one (because there is only one candidate). Furthermore, a maximum relative frequency (excluding dimension zero) at dimension K would be an indication for the need of (iii). Thus, any meaningful choice without going to (iii) should correspond to a peak of the relative frequency "in the middle". The smallest K that allows such a peak in the middle is 3 (Exercise 5.3). Now suppose that L gene-sets are bootstrapped. Usually it is not necessary to consider all possible dimensions up to L, especially in view of (iii). Note that the plot of the relative frequencies against all possible dimensions is typically

W-shaped with the peak in the middle occurring at a fairly low dimension. Here we consider all possible dimensions up to $[L/2]$, where $[x]$ is the largest integer $\leq x$. It follows that the minimum number of bootstrapped gene sets that allows a peak in the middle, when comparing dimensions up to $[L/2]$, is $L = 6$. The limited bootstrap allows one to focus only on the gene sets that are most "interesting". Statistically, this is equivalent to applying bootstrap to lower dimensional data.

Another issue occurs when the highest frequency occurs at dimension one. In this case, IF cannot tell whether it is one gene set, or no gene set (that is d.e.). Logically, the highest relative frequency does not constitute a "peak", if it occurs at dimension one, because the relative frequency at dimension zero is always 1 (a peak, by definition, is a relative frequency that occurs at a certain dimension, which is higher than the relative frequencies at adjacent dimensions). An important rule of the fence is called *conservative principle* ( ) which, in the IF case, says that whenever there are ties in the highest frequency, one should choose the highest dimension that ties for the highest frequency. Thus, if the highest relative frequency that occurs at dimension one is 1, by the conservative principle one chooses one gene set over no gene set. There is, however, an exception. What if the highest relative frequency occurs at dimension one, and it is less than 1? To solve this problem we assist IF with a test, called test for no gene set, when the situation occurs. The null hypothesis is that no gene set is d.e. The test statistic is the maximum of the restandardized maxmeans [Efron and Tibshirani (2007)] over all of the gene sets being considered (however, see Example 5.2). The critical value of the test is obtained by permutations. For example, suppose that there are 50 microarrays in the control and treatment groups, the controls being $1, \ldots, 50$ and treatments $51, \ldots, 100$. A random sample $i_1, \ldots, i_{100}$ is drawn without replacement from $1, \ldots, 100$ (i.e., a random permutation). Then, the new data matrix obtained by rearrange the columns of X according to $i_1, \ldots, i_{100}$. It is easy to see the rationale of the proposed test: If the null hypothesis is false, then at least one of the restandardized maxmeans is expected to be higher than the nominal level and so is the maximum of them. Note that, unlike the GSA test of Efron and Tibshirani (2007) which is for whether each individual gene set is significant, here the test is an overall assessment (whether some gene sets are significant, or no gene-set is significant).

Finally, there is an issue called *dominant factor*. We use an example for illustration. Suppose that two gene sets are d.e., of which the first is to a much greater extent than the second. The first gene set is then called a dominant factor. What happens is that the relative frequency of IF at dimension one tends to be (much) higher than that at dimension two, and hence results in underfitting. In other words, in this case, the IF tends to select the dominant factor and ignore the second gene set, even though it is also d.e. The dominant factor usually occurs when the sample size is limited—according to the asymptotic theory of IF (see Chapter 9), as $n \to \infty$, the relative frequency at dimension two goes to one (and that at any higher dimension stays strictly less than one) in this case, therefore, by the conservative principle, one would choose dimension two over dimension one. Nevertheless, our main concern is finite sample performance. Consider, again, the example. It is observed that, once the dominant factor is removed, the second gene-set begins to emerge. Therefore, a potential remedy is to apply IF, again, to the rest of the gene sets after the dominant factor is selected. In other words, the IF procedure is carried out in a sequential manner. Here, clearly, we need a stopping rule to avoid overfitting. Before a new round of IF is carried out we need to know whether any of the remaining gene sets is d.e., hence a test for no gene set, as described above, is performed. Still, there is a (small) chance that the test result is significant, even if no gene set is d.e., and this can happen at any round of IF. Therefore, theoretically, there is still a chance that the sequential procedure can go on and on, even if no gene sets is d.e. after the initial round. However, IF has provided us another useful information to stop the process when no gene sets is d.e. Recall the earlier discussion, where the idea is increase the number of bootstrapped gene sets until one finds a number, say, L, so that, when considering dimensions up to $[L/2]$, the highest frequency does not occur at dimension $[L/2]$. This indicates that the number of d.e. gene-sets is no more than $[L/2] - 1$, and hence sets up an upper bound. In other words, whatever one does, the total number of selected gene sets cannot exceed $[L/2] - 1$. We illustrate with an example.

Example 5.3. Consider the following hypothetical example. Suppose that $s_1 = 10$, $s_2 = s_3 = s_4 = 5$, and the rest of the s_i's are nearly zero. What is likely going to happen is that there are two peaks in the middle of the relative frequency plot (against the dimension). The first peak occurs at the dimension 1, corresponding to s_1, and the second at the dimension 4, corresponding to s_1, s_2, s_3, s_4. The heights of these peaks depend on how dominant s_1 is with respect to the rest of the gene sets (which we may not be able to tell with the given numbers). If s_1 is not so dominant, then the second peak will be higher; as a result, the subset $1, 2, 3, 4$ will be selected (which we suppose is the correct choice in this case). On the other hand, if

s_1 is dominant enough, the first peak will be higher; as a result, the subset 1 will be selected, but IF does not stop here. According to our procedure, following each selection (based on the relative frequencies) there is a test for no gene set. If the gene sets $2, 3, 4$ are strong enough, the null hypothesis (of no more gene set) will be rejected, so IF continues after gene set 1 is taken out. This time, the relative frequency plot is likely to have a single peak in the middle, and thus the subset $2, 3, 4$ will be selected. After the latest gene sets are taken out, the test for no gene set is likely to result in non-rejection of the null hypothesis; hence, the IF will stop and the selected gene sets are $1, 2, 3, 4$ (Exercise 5.4).

The rest of this section is devoted to a simulation study used to evaluate performance of the IF with comparison to the GSA of Efron and Tibshirani (2007). The latter authors carried out an empirical study, in which they simulated 1000 genes and 50 samples in each of 2 classes, control and treatment. The genes were evenly divided into 50 gene sets, with 20 genes in each gene set. The data matrix was originally generated independently from the $N(0, 1)$ distribution, then the treatment effect was added according to one of the following five scenarios:

1. All 20 genes of gene-set 1 are 0.2 units higher in class 2.
2. The first 15 genes of gene-set 1 are 0.3 units higher in class 2.
3. The first 10 genes of gene-set 1 are 0.4 units higher in class 2.
4. The first 5 genes of gene-set 1 are 0.6 units higher in class 2.
5. The first 10 genes of gene-set 1 are 0.4 units higher in class 2, and the second 10 genes of gene-set 1 are 0.4 units lower in class 2.

We consider the same five scenarios in our simulation study. In Efron and Tibshirani's study only the first gene set is of potential interest. We expand their one-gene-set case to a two-gene-set case, in which we duplicate the five scenarios to the second gene set.

Also, in Efron and Tibshirani's study the genes were simulated independently. We consider, in addition to the independent case ($\rho = 0$), a case where the genes are correlated with equal correlation coefficient $\rho = 0.3$. The correlation is generated by associating with each microarray a random effect. The genes on the same microarray are then correlated for sharing the same random effect. Let x_{ij} be the (i, j) element of the data matrix, X, where i represents the gene and j the microarray, $i = 1, \ldots, 1000$, $j = 1, \ldots, 100$. Here $j = 1, \ldots, 50$ correspond to the controls and $j = 51, \ldots, 100$ the treatments. Then, we have

$$x_{ij} = \alpha_j + \epsilon_{ij}, \tag{5.2}$$

where the α_j's and ϵ_{ij}'s are independent random effects and errors that are distributed as $N(0, \rho)$ and $N(0, 1 - \rho)$, respectively. It follows that each x_{ij} is distributed as $N(0, 1)$, and $\text{cor}(x_{ij}, x_{i'j}) = \rho$, $i \neq i'$. The treatment effects are then added to the right side of (5.2) for $j = 51, \ldots, 100$ and genes i in the given gene set(s), as above.

In Efron and Tibshirani's simulation study, the authors showed that the maxmean has the best overall performance as compared with other methods, including the mean, the absolute mean, GSEA [Gene Set Enrichment Analysis; Subramanian *et al.* (2005)] and GSEA version of the absolute mean. Therefore, we focus on the best performer of GSA, that is, the maxmean. In addition to the one-gene-set and two-gene-set cases, each with the five scenarios listed above, the simulation comparisons also include the case where no gene set is potentially interesting, that is, no treatment effect is added to any gene set. This is what we call the null scenario. For GSA one needs to choose the FDR as well as the number of permutation samples for the test of significance. For IF, on the other hand, one also needs to specify the level of significance as well as the number of permutation samples for the test for no gene set. The FDR and level of significance are both chosen as $\alpha = 0.05$. The number of permutations for both GSA and IF is 200 [which is the number that Efron and Tibshirani (2007) used in their simulation study].

The first comparison is on the probability of correct identification, or true-positive (TP). For IF this means that the gene sets selected match exactly those to which the treatment effects are added, which we call true gene-sets; similarly, for GSA this means that the gene sets that are found significant are exactly those true gene-sets. Table 5.1 reports the empirical probability of TP based on 100 simulation runs. For example, for the Null Scenario, One-Gene-Set case, with $\rho = 0$, the numbers mean that for 95 out of the 100 simulation runs, IF selected no (0) gene sets; while for 59 of the 100 simulation runs, GSA found no (0) gene sets. As another example, for Scenario 2, Two-Gene-Set case, with $\rho = 0.3$, IF selected the exact two gene sets, to which the treatment effects are added, for 97 out of the 100 simulation runs; while GSA found the exact two gene sets for 66 out of the 100 simulation runs. Note that these are results of same-data comparisons, that is, for each simulation run, the results for both methods are based on the same simulated data. Also reported (in the parentheses) are empirical probabilities of overfit (OF, in the sense that the identified gene sets include all the true gene-sets plus some false discoveries) and underfit (UF, in the sense that at least one of the true gene set is not discovered). It appears that IF has better performance than GSA in terms of TP uniformly across

Table 5.1 **IF vs GSA - Empirical Probabilities (in %) of TP (OF, UF).**

		$\rho = 0$		$\rho = 0.3$	
Scenario	Method	One-Gene-Set	Two-Gene-Set	One-Gene-Set	Two-Gene-Set
Null	IF	95 (5,0)	95 (5,0)	64 (36,0)	64 (36,0)
	GSA	59 (41,0)	59 (41,0)	52 (48,0)	52 (48,0)
1	IF	80 (6,14)	68 (1,31)	80 (16,4)	88 (4,8)
	GSA	53 (37,10)	53 (25,22)	61 (36,3)	62 (30,8)
2	IF	88 (5,7)	88 (0,12)	88 (12,0)	97 (2,1)
	GSA	67 (32,1)	65 (26,9)	65 (35,0)	66 (32,2)
3	IF	87 (5,8)	84 (0,16)	83 (14,3)	96 (2,2)
	GSA	66 (31,3)	68 (24,8)	66 (33,1)	69 (27,4)
4	IF	73 (6,21)	63 (2,35)	75 (15,10)	80 (7,13)
	GSA	64 (28,8)	57 (19,24)	66 (29,5)	63 (21,16)
5	IF	87 (6, 7)	84 (0,16)	91 (9,0)	99 (0,1)
	GSA	70 (30,0)	76 (17,7)	82 (18,0)	86 (12,2)

all the cases and scenarios. While most of the losses for IF are due to UF, OF appears to be the major problem for GSA. Furthermore, both methods appear to be fairly robust against correlations between genes.

Tables 5.2 and 5.3 report another set of summaries of the simulation results. Here reported are the mean numbers (over the simulation runs) of correctly identified gene sets (MC) and those of incorrectly identified gene-sets (MIC). The standard deviations for the mean numbers are also reported (in the parentheses; note that these are the standard deviations rather than the standard errors—the latter should be the s.d. divided by $\sqrt{100} = 10$, and therefore much smaller). For example, for Scenario 2, Two-Gene-Set case, with $\rho = 0.3$, the MC for IF is 1.99 (note that the true value is 2) with a s.d. of 0.10; the MIC for IF is 0.04 (there is no true value for MIC but, ideally, it should be 0) with a s.d. of 0.24. For GSA in this case, the MC is 1.98 with a s.d. of 0.14; the MIC is 0.35 with a s.d. of 0.56. It is seen that, for $\rho = 0$, GSA has higher MC but also higher MIC compared to IF. This is consistent with the observation from Table 1 that GSA tends to overfit while IF tends to underfit. Thus, in particular, GSA has a higher FDR compared to IF. On the other hand, the MC/MIC results are mixed for $\rho = 0.3$. Once again, there appear to be little difference between the case $\rho = 0$ and $\rho = 0.3$ for GSA. As for IF, the empirical MCs and MICs are both higher for $\rho = 0.3$, in most cases; however, the change does not seem to affect the overall performance.

Our next comparison focuses on consistency properties of both methods. Traditionally, consistency in model identification (including parameter estimation and model selection) involves sample size going to infinity. Such

Table 5.2 **IF vs GSA - Empirical MC (s.d.), MIC (s.d.):**
$\rho = 0$.

Scenario	Method	One-Gene-Set	Two-Gene-Set
Null	IF	0 (0), .06 (.27)	0 (0), .06 (.27)
	GSA	0 (0), .47 (.61)	0 (0), .47 (.61)
1	IF	.86 (.34), .06 (.23)	1.64 (.57), .03 (.17)
	GSA	.90 (.30), .42 (.57)	1.75 (.50), .33 (.53)
2	IF	.93 (.25), .05 (.21)	1.88 (.32), .01 (.10)
	GSA	.99 (.10), .36 (.55)	1.91 (.28), .30 (.50)
3	IF	.92 (.27), .05 (.21)	1.83 (.40), .01 (.10)
	GSA	.97 (.17), .35 (.55)	1.92 (.27), .26 (.46)
4	IF	.79 (.40), .06 (.23)	1.57 (.63), .03 (.17)
	GSA	.92 (.27), .35 (.55)	1.73 (.50), .25 (.45)
5	IF	.93 (.25), .06 (.23)	1.79 (.51), 01 (.10)
	GSA	1 (0), .33 (.53)	1.93 (.25), .19 (.41)

Table 5.3 **IF vs GSA - Empirical MC (s.d.), MIC (s.d.):**
$\rho = 0.3$.

Scenario	Method	One-Gene-Set	Two-Gene-Set
Null	IF	0 (0), .81 (1.17)	0 (0), .81 (1.17)
	GSA	0 (0), .57 (.66)	0 (0), .57 (.66)
1	IF	.96 (.19), .22 (.57)	1.92 (.27), .06 (.23)
	GSA	.97 (.17), .41 (.55)	1.92 (.27), .38 (.54)
2	IF	1 (0), .16 (.50)	1.99 (.10), .04 (.24)
	GSA	1 (0), .38 (.54)	1.98 (.14), .35 (.56)
3	IF	.97 (.17), .20 (.60)	1.98 (.14), .04 (.19)
	GSA	.99 (.10), .35 (.50)	1.96 (.19), .31 (.51)
4	IF	.90 (.30), .26 (.73)	1.87 (.33), .15 (.38)
	GSA	.95 (.21), .32 (.48)	1.83 (.40), .30 (.48)
5	IF	1 (0), .13 (.46)	1.99 (.10), 0 (0)
	GSA	1 (0), .20 (.45)	1.98 (.14), .12 (.32)

an assumption, however, is not very realistic in gene set analysis, because the sample size n is typically much smaller than the number of genes under consideration. Therefore, we consider a different type of consistency, called signal consistency. A gene-set identification procedure is signal-consistent if its probability of TP goes to one as the treatment effects, or signals, increase to infinity. Of course, one may not be able to increase the signals in real-life, but neither is the sample size going to infinity in real-life. The point is to see if a procedure works perfectly well in the "ideal situation", which we believe is a basic property, just like consistency in the traditional sense. We defer the detailed theoretical development to Chapter 9.

To investigate signal-consistency property of IF and GSA, we expand one of the cases, namely, the two-gene-set case of Scenario 5, by increasing

Table 5.4 **IF vs GSA - Empirical Probabilities (in %) of TP with Increasing Signals.**

Case #	Signals	$\rho = 0$		$\rho = 0.3$	
		IF	GSA	IF	GSA
1	(0.4,-0.4,0.4,-0.4)	84	76	99	86
2	(0.5,-0.5,0.5,-0.5)	100	88	100	97
3	(1.0,-1.0,1.0,-1.0)	100	100	100	100
4	(1.0,-1.0,0.5,-0.5)	100	97	100	99
5	(1.5,-1.5,0.5,-0.5)	100	88	100	88
6	(2.0,-2.0,0.5,-0.5)	100	64	100	56
7	(2.5,-2.5,0.5,-0.5)	100	26	100	23
8	(3.0,-3.0,0.5,-0.5)	100	10	100	3
9	(3.5,-3.5,0.5,-0.5)	100	2	100	0
10	(4.0,-4.0,0.5,-0.5)	100	0	100	0

the treatment effects in two different ways. First, we increase the signals in a balanced manner, that is, the signals increase at the same pace for both gene sets. Next, we let the signals increase in an unbalanced way, so that the pace is much faster for the first gene set than for the second. Table 5.4 reports the empirical probabilities of TP based on 100 simulation runs. Here the signals are expressed in the form of (a, b, c, d), where the values a, b, c, d are added to the right side of (5.2) for $51 \leq j \leq 100$ and 1st 10 genes of gene set one, 2nd 10 genes of gene set one, 1st 10 genes of gene set two, and 2nd 10 genes of gene set two, respectively. Case 1 is taken from the bottom two rows of Table 5.1 (two-gene-set case), which serves as a baseline. Then we see what happens when the signals increase. In cases 1, 2, 3, where the signals increase in the balanced way, both IF and GSA seem to work perfectly well as both methods show signs of signal-consistency. However, in cases 1, 4, ..., 10, where the signals increase in the unbalanced way, the empirical probability drops, and eventually falls apart for GSA, even with increasing signals. On the other hand, IF still shines in this situation, having perfect empirical probabilities of TP.

It is interesting to know what happens to GSA in the latest situation. The problem is restandardization. Efron and Tibshirani argued that re-standardization is potentially important in that it takes into account the overall distribution of the individual gene scores. Our simulation studies also confirmed that restandardization improves finite sample performance in some cases, not just for GSA but for IF as well (recall the initial ranking of the gene sets as well as the test for no gene set in IF are based on restandardized maxmeans). However, in the situation where the gene-set scores are dominated by, say, a single gene set, such as the above, the restandardized gene-set scores may look very different from those based on the permutation samples. Consider, for example, an extreme case where one gene set is so dominant that all but one gene-set scores are below the overall mean used in the restandardization. It follows that all but one re-standardized gene-set scores are negative. On the other hand, the critical value for any FDR that is commonly in use is expected to be, at least, non-negative. Therefore, a test based on comparing the restandardized gene-set scores with the critical value is expected to reject nothing but the null hypothesis corresponding to the dominant gene set, and ignore the potential interest of any others (even though some of them are actually d.e.).

In introducing the GSA method, Efron and Tibshirani (2007) considered a situation where the same treatment effect is added to all of the gene sets. In other words, all of the gene sets are equally d.e. The authors used this example to make the point for the need of restandardization. The claim is that, in this case, there is "nothing special about any one gene set". While the claim is arguable from a practical point of view, it would be interesting to see how the two methods, IF and GSA, work in a situation like this. Thus, as a final comparison, we simulated data according to Scenario 1 above, except that the 0.2 units are added to all of the gene sets. If, as the latest authors claimed, there is nothing special about any gene set, one expects a procedure to identify no (zero) gene set in this case. According to the results based on 100 simulation runs, when $\rho = 0$, the empirical probabilities of identifying zero gene set is 50% for IF and 47% for GSA; when $\rho = 0.3$, the corresponding empirical probabilities are 58% for IF and 54% for GSA. So, in the latest comparison, the two methods performed similarly with IF doing slightly better.

Finally, regarding computational efficiency of the methods compared to each other, it takes, for example, 4.0 second to run the IF for a single simulation under Scenario 2, One-Gene-Set case with $\rho = 0$ (see, for example, Table 5.1), as compared to 3.0 second to run the GSA for the same simulation, on a server computer [Intel(R) Core(TM)2 Extreme CPU X9650 @ 3.00GHz]. It should be noted that our simulation codes for IF are not written by a professional programmer. Nevertheless, in terms of the computational efficiency, IF is, at least, comparable to GSA.

5.4   Longitudinal study

Longitudinal studies were discussed in Section 4.2, where the focus was selection of the fixed covariates. On the other hand, a good model for variance-covariance (V-CV) structure of the longitudinal data is likely to improve quality of the inference, especially in terms of the standard errors and significance test. Furthermore, a model for the V-CV structure may be associated with the model for the fixed covariates. For example, consider the following.

Example 5.4. (Growth curve) A growth curve model for longitudinal data may be introduced as a two-level hierarchical model. Denote the longitudinal data by (y_{it}, x_{it}), where y_{it} is the response collected from the ith subject at time t, and x_{it} is a known covariate associated with the subject. In the first level, one has $y_{it} = \beta_{0i} + \beta_{1i}x_{it} + e_{it}$, where β_{0i} and β_{1i} are the intercept and slope that depend on the subject, and e_{it} is a random error that has mean 0 and variance σ_e^2. In the second level, one expresses the intercept and slope as $\beta_{0i} = \beta_0 + v_{0i}$ and $\beta_{1i} = \beta_1 + v_{1i}$, where β_0, β_1 are fixed, unknown coefficients, and $v_i = (v_{0i}, v_{1i})'$ is a bivariate random effect with mean vector zero, covariance matrix Σ, and is independent of e_{it}. If we combined the two levels, we have

$$y_{it} = \beta_0 + \beta_1 x_{it} + v_{0i} + v_{1i}x_{it} + e_{it}. \tag{5.3}$$

The mean of the response is $E(y_{it}) = \beta_0 + \beta_1 x_{it}$ (here x_{it} is considered fixed; in other words, the model is conditional on x_{it}); the variance of the response is $\operatorname{var}(y_{it}) = (1 \ x_{it})\Sigma(1 \ x_{it})' + \sigma_e^2$. It is clear that both the mean and variance depend on the same fixed covariate, x_{it}; therefore, selection of the fixed covariate will impact selection of the V-CV structure.

The example shows that the choice of the covariate should take into account the choice of the V-CV structure and vice versa. From this point of view, model selection for longitudinal data is complex.

In addition, in practice there are often many candidate variables for the fixed covariates, as shown in Section 4.4. In Section 4, the high-dimensional selection problem was tackled using the RF. Here we consider a different approach using the IF. To do so, the first thing we need is a subtractive measure. A natural choice is (5.1) with $s_j = |\hat{\beta}_j|$, where $\hat{\beta}_j$ is the estimate of the coefficient for the jth candidate covariate under the full model Mou (2012). This requires feasibility of fitting the full model, which is not always possible. For example, if the total number of candidate covariates is larger than the sample size, the LS fit is not possible; however, the lasso [Tibshirani (1996)] fit may provide a solution, which would shrink some of the β

coefficients to zero (see Chapter 7). Another good feature of this subtractive measure is that it is invariant under non-decreasing transformations. Namely, the measure with $s_j = |\hat{\beta}_j|$ is associated with the L^1 norm. There are, of course, other choices; for example, $s_j = \hat{\beta}_j^2$ would correspond to the L^2 norm, etc. However, the choice of the norm does not make a difference in view of the fast algorithm (see Section 5.2), because the ranking of the s_j's is the same regardless of the norm. Similar, one may use $s_j = h(|\hat{\beta}_j|)$, where $h(\cdot)$ is any strictly increasing function on $[0, \infty)$, and this makes no difference in terms of the ranking, and hence the selection result.

While it is appropriate to use IF for the selection of the fixed covariates, it may not be reasonable to use it for the selection of the V-CV structure. The reason is that the parameters involved in the latter may not present themselves in the way of "effects", or coefficients, of variables. For example, suppose that ρ is the correlation coefficient between the random effects, v_{0i} and v_{1i}, in (5.3). Note that $\rho = 0$ does not imply any of the random effects is absent. Due to such considerations, Mou (2012) proposed a two-stage selection procedure. In the first stage, the SAF (see Section 3.2) is applied to select the V-CV structure. Namely, let $y = (y_i)_{1 \leq i \leq m}$, where $y_i = (y_{it})_{t \in T_i}$, T_i being the set of observational times for subject i, and m is the total number of subjects. Let A be an $N \times (N - p_f)$ matrix, where $N = \sum_{i=1}^{m} n_i$ with n_i being the cardinality of T_i, and $p_f = \text{rank}(X_f)$, X_f being the full matrix of covariates, whose columns correspond to all of the fixed covariates, such that $A'X_f = 0$. The SAF is applied to $z = A'y$, whose distribution is free of the models for the fixed covariates, to select the V-CV model. The transformation by A is similar to that used in RF (see Section 4), motivated by the REML [e.g., Jiang (2007)]. Given the selected model, in the second stage, the fixed covariates are selected using the IF.

Mou (2012) carried out a simulation study on the performance of the two-stage procedure under the following linear mixed model for longitudinal data [e.g., Diggle *et al.* (2002), sec. 5.1]:

$$y_{ij} = x_{ij}'\beta + v_i + W_i(t_{ij}) + e_{ij}, \tag{5.4}$$

where y_{ij} is the response from subject i at the jth observational time for the subject, t_{ij}; and x_{ij} is a vector of covariates whose first component is 1, corresponding to the intercept. Furthermore, v_i is a subject-specific random effect with mean 0 and variance σ_0^2, $W_i(\cdot)$ is a zero-mean stationary Gaussian process with variance σ_1^2 and correlation function $\rho(\cdot)$; and e_{ij} is a Gaussian random error with mean 0 and variance σ_2^2. Let n_i be the total number of responses for subject i. It is assumed that $v_i, W_i(\cdot), 1 \leq i \leq m$

are independent, $e_{ij}, 1 \leq j \leq n_i, 1 \leq i \leq m$ are independent, and the e_{ij}'s are independent with the v_i's and $W_i(\cdot)$'s. Let ϵ_{ij} denote the right side of (5.3) without $x'_{ij}\beta$, and $\epsilon_i = (\epsilon_{ij})_{1 \leq j \leq n_i}$. It can be shown (Exercise 5.5) that the covariance matrix of ϵ_i has the expression

$$\text{Var}(\epsilon_i) = \sigma_0^2 J_i + \sigma_1^2 H_i + \sigma_2^2 I_i, \tag{5.5}$$

where I_i, J_i are the $n_i \times n_i$ identity matrix and matrix of 1's, respectively, and $H_i = (h_{ijk})_{1 \leq j,k \leq n_i}$ with $h_{ijk} = \rho(|t_{ij} - t_{ik}|)$, the correlation coefficient between $W_i(t_{ij})$ and $W_i(t_{ik})$. Furthermore, because $\epsilon_1, \ldots, \epsilon_m$ are independent, the V-CV structure of the data is block-diagonal with the diagonal blocks given by (5.5).

As for the Gaussian process $W_i(\cdot)$, in the simulation study, the author has focused on the autoregressive (AR) process [e.g., Shumway (1988)]. The latter is defined as a random process $W_t, t \in Z = \{0, \pm 1, \pm 2, \ldots\}$ satisfying

$$W_t = a_1 W_{t-1} + \cdots + a_p W_{t-p} + \eta_t, \tag{5.6}$$

where η_t, $t \in Z$, is a white noise process such that $\text{E}(\eta_t) = 0$, $\text{var}(\eta_t) = \tau^2$ ($= 1$ in the simulation study), and $\text{cov}(\eta_s, \eta_t) = 0$ for $s \neq t$, the a's are unknown parameters, and p is the order of the AR. The AR process is stationary if the corresponding polynomial $A(z) = \sum_{j=0}^{p} a_j z^j \neq 0$ for all complex z with $|z| \leq 1$. The model is denoted by AR(p).

The following models are considered for the V-CV structure:

Model 1: Independent errors, that is, (5.4) without v_i and $W_i(t_{ij})$.

Model 2: Linear mixed model only, that is, (5.4) without $W_i(t_{ij})$.

Model 3: (5.4) with $W_i(\cdot)$ being AR(1).

Model 4: (5.4) with $W_i(\cdot)$ being AR(2).

Model 5: (5.4) with $W_i(\cdot)$ being AR(3).

Model 6: (5.4) with $W_i(\cdot)$ being AR(4).

In all cases and scenarios considered below, equally spaced time points are considered. First consider the situation where both the random effects and the AR processes are normally distributed.

Case I: Scenario 1: m, the number of subjects, is 30; $n_i = 5, 1 \leq i \leq m$; the true β is $(2, 10, 0, 0, 8, 0, 0, 7, 0)$; the true V-CV model is Model 3 with $\sigma_0^2 = 1$, $a_1 = 0.15$, and $\sigma_r^2 = 1, r = 1, 2$. Scenario 2: Same as Scenario 1 except that m is increased to 60. Scenario 3: Same as Scenario 1 except that $m = 100$ and $n_i = 10$.

Case II: Scenario 1: Same as Case I, Scenario 1 except that the true β is $(2, 0, 0, 0, 0, 0, 0, 7, 0)$, that is, the true model for the fixed covariates is lower-dimensional. Scenarios 2 and 3: Same as Case I, Scenarios 2 and 3, respectively, except with the new true β.

Table 5.5 **Case I.**

Scen.	Method	TP (%)	UF (%)		OF (%)		MC (s.d.)	MIC (s.d.)
			V-CV	FC	V-CV	FC		
1	TSF	21.5	75.0	0.0	3.5	0.0	4.0 (0.0)	0.0 (0.0)
	AIC	19.0	57.0	0.0	14.5	33.5	4.0 (0.0)	0.4 (0.6)
	BIC	16.0	75.5	0.0	1.5	33.5	4.0 (0.0)	0.4 (0.6)
2	TSF	20.0	70.5	0.0	9.5	0.0	4.0 (0.0)	0.0 (0.0)
	AIC	15.5	58.5	0.0	6.5	30.0	4.0 (0.0)	1.2 (0.5)
	BIC	7.0	87.5	0.0	1.0	32.0	4.0 (0.0)	1.2 (0.5)
3	TSF	93.0	1.0	0.0	6.0	0.0	4.0 (0.0)	0.0 (0.0)
	AIC	73.0	0.0	0.0	17.0	11.0	4.0 (0.0)	0.1 (0.3)
	BIC	85.5	2.0	0.0	3.5	11.0	4.0 (0.0)	0.1 (0.3)

Case III: Scenario 1: Same as Case I, Scenario 1 except that the true β is $(2, 10, 5, 0, 8, 0, 6, 7, 0)$, that is, the true model for the fixed covariates is lower-dimensional, and that the true V-CV model is Model 4 with $a_1 = 0.15$, $a_2 = 0.65$. Scenario 2: Same as Scenario 1 except with $m = 60$. Scenario 3: Same as Scenario 1 except with $m = 100, n_i = 10$.

The results are reported in Tables 5.5–5.6. Here the TP is for the combined model of both fixed covariates (FC) and V-CV structure; the UF and OF are for the FC and V-CV models seperately, and the MC and MIC are for the FC model only. The two-stage (fence) procedure is denoted by TSF. The results are compared with the AIC and BIC procedures, based on the same-data comparisons of 200 simulation runs. What is most impressive is that the IF never missed, maintaining a perfect record out of all of the simulation runs, as indicated by these tables (both UF and OF are exactly zero, so is the MIC, and the MC is exactly the number of non-zero β coefficients). It follows that all of the misses by TSF are due to the misidentification of the V-CV structure. Overall, TSF outperforms AIC and BIC, especially when the sample size is relatively large. Between AIC and BIC, the former performs better under the smaller sample size, but the trend is reversed under the larger sample size. The trend is not surprising. It is well known that AIC is inconsistent [Shibata (1984)] while BIC is consistent [Hannan (1980)] under the classical setting of model selection, where the space of candidate models is finite, with the true model being a member.

Another situation is also considered in the simulation study, in which the random effects are generated from some non-Gaussian distributions (the AR process is still Gaussian). The set-up is the following.

Case IV: $m = 100$, $n_i = 10$; the true β is the same as in Case I; and the V-CV model is the same as in Case III. Three scenarios are considered: Scenaria A: $v_i \sim t_3$; Scenario B: v_i has the same distribution as $\xi - 1$, where

Table 5.6 **Case II.**

Scen.	Method	TP (%)	UF (%)		OF (%)		MC (s.d.)	MIC (s.d.)
			V-CV	FC	V-CV	FC		
1	TSF	19.0	74.0	0.0	7.0	0.0	2.0 (0.0)	0.0 (0.0)
	AIC	3.0	88.0	0.0	2.0	63.5	2.0 (0.0)	1.5 (0.7)
	BIC	5.5	67.0	0.0	17.0	64.0	2.0 (0.0)	1.5 (0.7)
2	TSF	23.0	66.5	0.0	6.0	0.0	2.0 (0.0)	0.0 (0.0)
	AIC	10.5	67.0	0.0	12.5	44.0	2.0 (0.0)	1.3 (0.5)
	BIC	8.0	85.0	0.0	2.0	42.5	2.0 (0.0)	1.3 (0.6)
3	TSF	49.0	41.5	0.0	9.5	0.0	2.0 (0.0)	0.0 (0.0)
	AIC	48.5	24.5	0.0	19.0	17.0	2.0 (0.0)	1.1 (0.3)
	BIC	36.0	56.5	0.0	1.5	18.5	2.0 (0.0)	1.1 (0.3)

Table 5.7 **Case III.**

Scen.	Method	TP (%)	UF (%)		OF (%)		MC (s.d.)	MIC (s.d.)
			V-CV	FC	V-CV	FC		
1	TSF	11.5	85.5	0.0	3.0	0.0	6.0 (0.0)	0.0 (0.0)
	AIC	21.5	52.5	0.0	13.0	35.0	6.0 (0.0)	1.2 (0.4)
	BIC	12.5	73.0	0.0	7.0	35.0	6.0 (0.0)	1.2 (0.4)
2	TSF	51.0	46.5	0.0	2.5	0.0	6.0 (0.0)	0.0 (0.0)
	AIC	41.5	26.0	0.0	18.0	21.5	6.0 (0.0)	1.2 (0.4)
	BIC	32.0	53.0	0.0	2.0	21.5	6.0 (0.0)	1.2 (0.5)
3	TSF	81.5	0.0	0.0	18.5	0.0	6.0 (0.0)	0.0 (0.0)
	AIC	35.0	0.0	0.0	61.0	12.5	6.0 (0.0)	1.0 (0.0)
	BIC	77.5	0.0	0.0	13.5	11.0	6.0 (0.0)	1.0 (0.0)

Table 5.8 **Case IV.**

Scen.	Method	TP (%)	UF (%)		OF (%)		MC (s.d.)	MIC (s.d.)
			V-CV	FC	V-CV	FC		
1	TSF	76.0	0.0	0.0	24.0	0.0	4.0 (0.0)	0.0 (0.0)
	AIC	29.5	0.0	0.0	62.0	19.5	4.0 (0.0)	1.1 (0.3)
	BIC	64.5	0.0	0.0	19.0	20.0	4.0 (0.0)	1.1 (0.3)
2	TSF	82.0	18.0	0.0	0.0	0.0	4.0 (0.0)	0.0 (0.0)
	AIC	35.5	0.0	0.0	60.0	15.0	4.0 (0.0)	1.1 (0.3)
	BIC	66.5	0.0	0.0	20.5	16.0	4.0 (0.0)	1.0 (0.2)
3	TSF	79.5	20.5	0.0	0.0	0.0	4.0 (0.0)	0.0 (0.0)
	AIC	35.5	0.0	0.0	56.0	19.5	4.0 (0.0)	1.1 (0.3)
	BIC	66.5	0.0	0.0	16.0	19.5	4.0 (0.0)	1.1 (0.3)

$\xi \sim$ Exponential(1); Scenario C: v_i has the normal-mixture distribution with equal weights given to $N(-1, 0.25)$ and $N(1, 0.25)$. The results, again based on 200 simulation runs, are reported in Table 5.8. The overall pattern is very much the same as in the normal case. In particular, the IF still has a perfect performance out of all of the simulation runs.

A real data application is considered in Section 5.6.

5.5   Relative IF

The IF makes its choice of the optimal dimension by comparing the highest selection frequencies at different dimensions. In a way, this may not seem entirely "fair". For example, suppose that there are n candidate variables. It might seem a lot easier for one variable to "stand out" among all of the n variables than for a pair of variables to stand out among all pairs of the n variables (just counting the numbers, there are $n-1$ "competitors to beat" in the former case, and $n(n-1)/2 - 1$ competitors in the latter).

One way to make the comparison more fair is to compare relative significance, or "strangeness", of the selection frequencies rather than their actual values. This is similar to the p-value in hypothesis testing [e.g., Lehmann (1986)]. A main step is to construct an reference distribution for a particular application. We use the term *variables* that, in practice, correspond to variables, factors, or gene sets. Suppose that there are a total of n candidate variables, numbered as $1, \ldots, n$, and we are interested in selecting the best subset of up to q variables, where $1 \le q \le n$. We shall assume that at least one variable is of interest, in reality. Thus, the new method is more restrictive than IF, which can also deal with the case in which no variable is of interest, in reality. On the other hand, the assumption is not unreasonable. In practice, there are often indications from other sources suggesting that some variables among the candidates are of interest; therefore, we can focus on the problem of identify these variables without having to worry about their existance.

Let $M(B, n, p)$ denote a random $B \times n$ matrix whose rows are independent. Furthermore, p components of each row of $M(B, n, p)$ are randomly chosen to be assign the value 1, and the rest of the elements of $M(B, n, p)$ (that are not assigned the value 1) are 0. The motivation is that each row of $M(B, n, p)$ corresponds to an indication of the top p variables that are randomly selected, and this process is repeated independently B times. Let $M(B, n, p)_i$ denote the ith column of $M(B, n, p)$. Define $c(i, B, n, 1) = 1'_B M(B, n, 1)_i$, $c(i, j, B, n, 2) = 1'_B\{M(B, n, 2)_i \cdot M(B, n, 2)_j\}$, $c(i, j, k, B, n, 3) = 1'_B\{M(B, n, 3)_i \cdot M(B, n, 3)_j \cdot M(B, n, 3)_k\}$, and so on, for $i < j < k < \cdots$, where 1_B denotes the $B \times 1$ vector of 1's, and $a \cdot b$ is the Hadamard product between two vectors, $a = (a_1, \ldots, a_B)'$ and $b = (b_1, \ldots, b_B)'$, such that $a \cdot b = (a_1 b_1, \ldots, a_B b_B)'$. Intuitively, $c(i_1, \ldots, i_p, B, n, p)$ is the count of the times that the indexes $i_1 < \cdots < i_p$ are selected as the top p variables among the B repeated experiments. We then compute the maximum frequecies

$$f(B, n, 1) = B^{-1} \max_{1 \leq i \leq n} c(i, B, n, 1),$$

$$f(B, n, 2) = B^{-1} \max_{1 \leq i < j \leq n} c(i, j, n, 2),$$

$$f(B, n, 3) = B^{-1} \max_{1 \leq i < j < k \leq n} c(i, j, k, B, n, 3), \tag{5.7}$$

and so on. The distribution of $f(B, n, p)$, denoted by $F(B, n, p)$, is what we call a reference distribution.

In IF, suppose that the maximum frequencies $f_p^*, 1 \leq p \leq q$ are obtained from the data based on B bootstrap samples. We propose the following additional step by comparng f_p^* to the corresponding reference distribution to obtain the p-value of f_p^*. The precise definition of the p-value is given below. The optimal dimension of the model, which is all that is needed under the IF, is the p whose corresponding f_p^* has the smallest p-value. By considering the p-value of f_p^* instead of f_p^* itself, we are comparing the different dimensions in their relative "strangeness" with respect to the reference distribution. For example, it is likely that the quantiles of $F(B, n, 2)$ are relatively smaller compared to the corresponding quantitles of $F(B, n, 1)$, making it easier for f_2^* to receive a smaller p-value than f_1^*.

The construction of the reference distribution requires computation of the probabilities under the distribution. Consider a process of drawing m items, with replacement, from K items. The number of repeated appearance of a given item among the m draws is called the frequency of the item, and the maximum of the frequencies of all different items in the draws is called the maximum frequency, denoted by $X_{K,m}$. It It follows that, for any possible value of $X_{K,m}$, we have

$$P(X_{K,m} = i) = \frac{N(K, m, i)}{K^m}, \tag{5.8}$$

where $N(K, m, i)$ is the number of cases that $X_{K,m} = i$. To see how (5.8) is related to the reference distribution, note that if the selection of the p out of n variables are completely random, each of the $K = C_p^n$ such selections are equally likely, where C_p^n is the combination number of choosing p items out of n items. This is equivalent to drawing B items with replacement from the K items. Thus, by (5.8), we have

$$P\{f(B, n, p) = i\} = (C_p^n)^{-B} N\left(C_p^n, B, i\right), \tag{5.9}$$

and $F(B, n, p)(x) = \sum_{i \leq x} P\{f(B, n, p) = i\}$. Given the observed maximum relative frequency, $f_p^* = x/B$, where $x \in \{1, \ldots, B\}$, the corresponding p-value is $1 - F(B, n, p)(x - 1) = \sum_{i \geq x} P\{f(B, n, p) = i\}$. Therefore, all we have to do is to compute $N(K, m, i)$, and we shall develope a recursive formula to do the computation.

It is easy to show that

$$N(K, m, 1) = 1_{(m \leq K)} K(K - 1) \cdots (K - m + 1). \tag{5.10}$$

If $i > 1$, then $N(K, m, i) = 0$ if $m > Ki$. If $m \leq Ki$, then $N(K, m, i)$

= the number of cases that there is at least one i-group, meaning that an item is repeated i times, and no $(i + 1)$-group

$= \sum_{1 \leq s \leq m/i}$ the number of cases that there are exactly s non-overlaping i-groups; the rest not overlapping with the s groups and have maximum frequency $\leq i - 1$

$= \sum_{1 \leq s \leq m/i} \sum_s \sum_{x_1, \dots, x_s} \#(s; x_1, \dots, x_s; i - 1)$,

where $\sum_s$ denotes the summation over all ways of forming the s groups; order does not matter, $\sum_{x_1, \dots, x_s}$ the summation over all ways of assigning the s groups with different items, $x_1, \dots, x_s \in \{1, \dots, K\}$, and $\#(s; x_1, \dots, x_s; i - 1) =$ the number of cases such that the first group $= x_1$, $\dots$, the sth group $= x_s$; the rest $\in \{1 \dots, K\} \setminus \{x_1, \dots, x_s\}$ and have maximum frequency $\leq i - 1$.

It is easy to show that the number of terms involved in $\sum_s$ is equal to

$$\frac{C_i^m C_i^{m-i} \cdots C_i^{m-si+i}}{s!} = \frac{m!}{(i!)^s (m - si)! s!}.$$

Meanwhile, the number of terms involved in $\sum_{x_1, \dots, x_s}$ is equal to $K(K - 1) \cdots (K - s + 1)$. Furthermore, it is easy to show that $\#(s; x_1, \dots, x_s; i - 1) = 1$ if $K = s$ (which implies $m = Ki$) or $K > s, m = si$; otherwise, $\#(s; x_1, \dots, x_s; i - 1) = \sum_{j=1}^{i-1}$ the number of ways of drawing $m - si$ items from $K - s$ items with replacement such that the maximum frequency is j (note that $m > si$ implies $K > s$) $= \sum_{j=1}^{i-1} N(K - s, m - si, j)$. From the above derivation, we obtain the following recursive formula:

$$N(K, m, i)$$
$$= 1_{(m \leq Ki)} m! \sum_{1 \leq s \leq m/i} \frac{K(K - 1) \cdots (K - s + 1)}{(i!)^s (m - si)! s!} \{1_{(K=s \text{ or } K>s, m=si)}$$
$$+ 1_{(m>si)} \sum_{j=1}^{i-1} N(K - s, m - si, j)\}. \tag{5.11}$$

Notes. Although (5.11) is derived by assuming that $i > 1$, it actually also holds for $i = 1$, according to (5.10) (see below). Also, it might be thought that something equivalent to (5.11) could be found in the literature of, say, combinatorial mathematics; however, we are unable to find such a result. On the other hand, see below for an alternative computational tool. We verify (5.11) for some simple cases.

Example 5.5. As our first example, we show that (5.10) is a special case of (5.11), with $i = 1$, even though (5.11) is derived assuming $i > 1$. Define $\sum_{j=1}^{0} \cdots = 0$. It is easy to show that $1_{(K=s \text{ or } K>s,m=s)} = 1_{(s=m)}$, if $m \leq K$. Thus, from (5.11), we have $N(K, m, 1) = 1_{(m \leq K)} m! K(K - 1) \cdots (K - m + 1)/(1!)^m (m - m)! m! = 1_{(m \leq K)} K(K - 1) \cdots (K - m + 1)$, which is (5.10).

Example 5.6. As our next example, consider the case $K = 2, m = 1$. In this case, by directly counting, it is easy to see that $N(2, 1, 1) = 2$; and $N(2, 1, i) = 0, i > 1$. On the other hand, by (5.11), we have $N(2, 1, 1) = 1_{(1 \leq 2)} 2 = 2$; and $N(2, 1, i) = 1_{(1 \leq 2i)} 1! \sum_{1 \leq s \leq 1/i} \cdots = 0, i > 1$. Thus, (5.11) is verified.

Example 5.7. Next, consider the case $K = m = 2$. Again, by directly counting, it is easy to see that $N(2, 2, 1) = N(2, 2, 2) = 2$. On the other hand, by (5.11), we have $N(2, 2, 1) = 1_{(2 \leq 2)} 2(2 - 2 + 1) = 2$; $N(2, 2, 2) = 1_{2 \leq 2 \cdot 2} 2! \sum_{1 \leq s \leq 2/2} \cdots = 2\{2/(2!)^1 (2-2)! 1!\} \{1_{(2=1 \text{ or } 2>1,2=2)} + 1_{(2>2)} \sum_{j=1}^{1} \cdots \} = 2$. Thus, (5.11) is verified for $i = 1, 2$.

Our final example is left as an exercise (Exercise 5.6).

Although computer codes based on (5.11) can be easily written to implement the computation under the reference distribution, we have found an alternative tool that is available in the birthday package in R, namely, pbirthday(). A reference for the R package is Diaconis and Mosteller (1989). For example, in pbirthday(n, classes=365, coincident=2), n is *experimental times*, which corresponds to B, the number of bootstrap samples in the gene set selection problem; classes is the number of different birthdays (in a non-leap year), which corresponds to the number of different gene sets; and coincident is *repeating times*, that is, two people having the same birthday, which corresponds to the maximum frequency. In addition to Examples 5.5–5.7 (and Exercise 5.6), formula (5.11) was verified numerically for a number of other cases with larger K, m, and i using the birthday package.

To see how the RIF idea works, we compare the performance of RIF with that of the IF in a simulation study under the same setting as in Table 5.1 but without the null scenario. As noted, a testing procedure is incorporated with the IF to overcome the dominant factor effect (see Section 5.3, the paragraph above Example 5.3). In a way, the RIF is also designed to deal with the dominant factor. Thus, we compare the RIF with IF both

Table 5.9 **RIF vs IF: Empirical Probability (in %) of TP (OF, UF).**

Scenario	Method	$\rho = 0$		$\rho = 0.3$	
		One-Gene-Set	Two-Gene-Set	One-Gene-Set	Two-Gene-Set
1	RIF	79 (16, 5)	70 (3, 27)	87 (3, 10)	80 (3, 17)
	IF	92 (2, 6)	35 (1, 64)	89 (0, 11)	53 (2, 45)
2	RIF	96 (4, 0)	94 (1, 5)	92 (3, 5)	96 (2, 2)
	IF	100 (0, 0)	64 (0, 36)	95 (0, 5)	76 (0, 24)
3	RIF	96 (4, 0)	87 (3, 10)	92 (3, 5)	93 (0, 7)
	IF	99 (0, 1)	56 (0, 44)	95 (0, 5)	64 (0, 36)
4	RIF	84 (13, 3)	65 (9, 26)	89 (5, 6)	87 (1, 12)
	IF	96 (1, 3)	30 (0, 70)	94 (0, 6)	54 (0, 46)
5	RIF	85 (15, 0)	82 (9, 9)	96 (4, 0)	95 (3, 2)
	IF	100 (0, 0)	45 (1, 54)	100 (0,0)	80 (0, 20)

without the testing procedure. The results, based on 100 simulation runs, are presented in Table 5.9. Reported are empirical probabilities of TP (OF-overfit, UF-underfit). It is seen that, while IF performs slightly better in the one-gene-set case, RIF performs much better in the two-gene-set case, indicating that RIF is less conservative in picking up more gene sets. A further look shows that most of the misidentifications of IF in the two-gene-set case are due to UF, suggesting that the RIF idea is successful in overcoming the dominant factor.

5.6 Real data examples

5.6.1 *The p53 data*

The p53 NCI-60 data [Subramanian *et al.* (2005)], mentioned in Section 5.2, consists of a total of 522 functional gene sets. Gene expression values are measured on 17 cell line samples labeled as normal, and 33 samples labeled as mutated. As noted, a question of biological interest is to identify gene pathways (i.e., groups of genes acting in concert with one another) that are differentially expressed under different conditions. The data were previously analyzed by Subramanian *et al.* (2005) using the gene set enrichment analysis (GSEA), and Efron and Tibshirani (2007) using the GSA.

We apply both IF and RIF to this data. There are a total of 10,100 genes, some of which are not involved in any pathways, which we call gene sets. The original 522 gene sets include 5 duplicated ones. The number of genes in those gene sets range from 2 to 447. An initial "clean-up" was carried out before the IF/RIF analysis, in which we exclude the duplicated gene sets as well as gene sets that contain less than 15 genes–there are a total

Table 5.10 **Analysis of p53 Data: Gene Sets Identified by Different Methods (X Indicating Identification; * Indicating Non-identification).**

IF/RIF	GSEA	GSA
ctlPathway	*	*
hsp27Pathway	X	X
il17Pathway	*	*
no2il12Pathway	*	*
p53hypoxiaPathway	X	X
p53Pathway	X	X
radiation_sensiticity	X	X
SA_G1_AND_S_ PHASES	*	X
p53_UP	X	X
*	rasPathway	X
*	*	n.g.f.Pathway

of 194 such gene sets, as was also done by Efron and Tibshirani (2007). The IF and RIF have identified the same gene sets, as reported in Table 5.10. As a comparison, the gene sets identified by the GSEA [Subramanian *et al.* (2005)] and GSA [Efron and Tibshirani (2007)] are also reported. Subramanian *et al.* (2005) reported the GSEA results with the FDR cut-off level 0.25; Efron and Tibshirani (2007) reported the GSA results with the FDR cut-off level 0.10. There are, of course, some overlaps and non-overlaps, but, notably, the five gene sets identified by all of the methods (hsp27Pathway, p53hypoxiaPathway, p53Pathway, radiation_sensitivity, p53_UP) are among the six gene sets marked in Table 1 of Efron and Tibshirani (2007) as "significant". The only gene set listed as "significant" in Efron and Tibshirani (2007) but not identified by IF/RIF is rasPathway. Also, the gene set SA_G1_AND_S_PHASES is identified by IF/RIF and GSA but not by GSEA; the gene set n.g.f.Pathway is identified by GSA but not by IF/RIF or GSEA. Finally, IF/RIF have identified three gene sets (ctlPathway, il17Pathway, no2il12Pathway) that are neither identified by GSEA nor by GSA.

5.6.2 *The milk data*

We use a real data example to illustrate the TSF method discussed in Section 5.4. Diggle *et al.* (2002) discussed data collected weekly from 79 Australia cows. The response variable was protein content of milk. The cows were fed on one of three diets: barley, a mixture of barley and lupins, or lupins alone. More specifically, there were 25 cows in the barley group, 27 in the mixture group, and 27 in the lupins group. The repeated measures on each cow, up to 19 weekly samples, were joined to accentuate the longitudinal nature of the data. The objective of the study was to determine how diet affects the protein content.

The spaghetti plots, presented in Fig. 1.4 of Diggle *et al.* (2002), show a trend of protein content initially going down linearly and eventually picking up somewhat nonlinearly. Based on such observations, Diggle *et al.* (2002) proposed the following model for the mean response. Let $x_1 = 1$ if diet = barley, and 0 otherwise; $x_2 = 1$ if diet = barley, -1 if diet = mixture, and 0 otherwise; $x_3 = x_2 t$, where t represents the time (week); $x_4 = 1$ if diet = barley, -1 if diet = lupins, and 0 otherwise; $x_5 = x_4 t$; $x_6 = t$ if $t \leq 3$, and 3 if $t > 3$; $x_7 = 0$ if $t \leq 3$, and $t - 3$ if $t > 3$; and $x_8 = 0$ if $t \leq 3$, and $(t - 3)^2$ if $t > 3$. The mean function for the three diet groups (1 for barley, 2 for mixture, 3 for lupins) are

$$\mu_1(t) = \begin{cases} \beta_{01} x_1 + \beta_1 x_6, & \text{if } t \leq 3, \\ \beta_{01} x_1 + \beta_1 x_6 + \beta_2 x_7 + \beta_3 x_8, & \text{if } t > 3; \end{cases}$$

$$\mu_2(t) = \begin{cases} \beta_{021} x_2 + \beta_{022} x_3 + \beta_1 x_6, & \text{if } t \leq 3, \\ \beta_{021} x_2 + \beta_{022} x_3 + \beta_1 x_6 + \beta_2 x_7 + \beta_3 x_8, & \text{if } t > 3; \end{cases}$$

$$\mu_3(t) = \begin{cases} \beta_{031} x_4 + \beta_{032} x_5 + \beta_1 x_6, & \text{if } t \leq 3, \\ \beta_{031} x_4 + \beta_{032} x_5 + \beta_1 x_6 + \beta_2 x_7 + \beta_3 x_8, & \text{if } t > 3. \end{cases}$$

If we define $X_s = x_1 1_{(r=1)}$, $X_2 = x_2 1_{(r=2)}$, $X_3 = x_3 1_{(r=2)}$, $X_4 = x_4 1_{(r=3)}$, $X_5 = x_5 1_{(r=3)}$, $X_6 = x_6$, $X_7 = x_7 1_{(t>3)}$, and $X_8 = x_8 1_{(t>3)}$, then, we have a unified expression (Exercise 5.7),

$$\mu_r(t) = \beta_{01} X_1 + \beta_{021} X_2 + \beta_{022} X_3 + \beta_{031} X_4 + \beta_{032} X_5$$
$$+ \beta_1 X_6 + \beta_2 X_7 + \beta_3 X_8. \tag{5.12}$$

(5.12) may be considered as a full model for the mean function. It would be interesting to see if the model can be simplified. Mou (2012) considered selection among the fixed covariates, $X_1, \ldots, X_8$. In addition to the mean function, several models for the V-CV structure were considered. These models correspond to the V-CV Models 1–6 in Section 5.4. The first stage of TSF selects the V-CV model. Figure 5.1 shows the AF frequency plot vs c. The highest peak in the middle occurs at $c^* = 41$. With $c = c^*$, the fence has selected Model 3, which has all three error terms on the right side of (5.4), v_i, $W_i(t_{ij})$, and e_{it}, with $W_i(\cdot)$ being the AR(1) process. Given the selected V-CV model, in the second stage, the IF is applied to select the fixed covariates. The optimal model for the fixed covariates involves X_1, X_2, X_4 and X_6, which corrsponds to the mean function

$$\mu_r(t) = \begin{cases} \beta_{0r} + \beta_1 t, & \text{if } t \leq 3, \\ \beta_{0r} + 3\beta_1, & \text{if } t > 3 \end{cases} \tag{5.13}$$

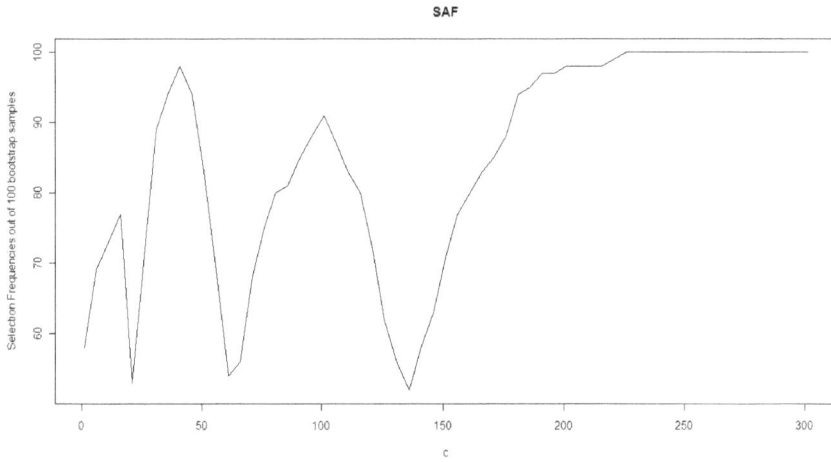

Fig. 5.1   *The SAF Frequency Plot*

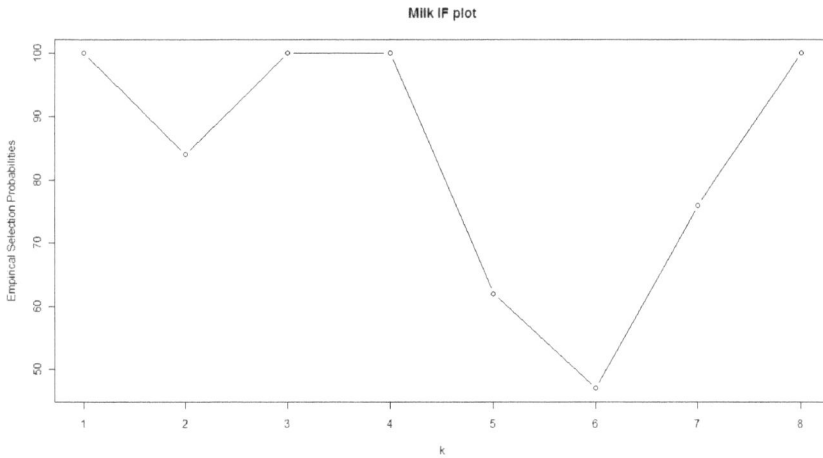

Fig. 5.2   *The IF Frequency Plot*

for $r = 1, 2, 3$ (Exercise 5.7). Figure 5.2 displays the IF frequency plot. As can be seen, two dimensions have tied for the highest frequency in the middle, $d = 3$ and $d = 4$. Thus, by the conservative principle (see Section 5.3), $d = 4$ is selected, which corresponds to the AR(1).

5.7   Exercises

5.1. In this exercise, you are asked to compare computational efficiency associated with a subtractive measure with a non-subtractive measure using simulated data.

a.   Generate data under the following linear regression model: $y_i = x_i'\beta + \epsilon_i$, $i = 1, \ldots, 100$, where $x_i = (1, x_{i,1}, \ldots, x_{i,24})'$, and the components $x_{i,j}, 1 \leq j \leq 24$ are generated from the $N(0,1)$ distribution; so are the ϵ_i's. Furthermore, the first four components of β are 1; the rest are 0 (so the true model has dimension 4).

b.   Consider two measures of lack-of-fit, Q_1 and Q_2. Q_1 is given by (5.1) with $s_j = |\hat{\beta}_j|$, where $\hat{\beta}_j$ is the least squares (LS) estimate of the jth component of β, obtained by fitting the full model, and $s = \sum_{j=1}^{25} s_j$. Q_2 is the RSS defined by $Q_2(M) = \sum_{i=1}^{100}(y_i - x_{M,i}'\hat{\beta}_M)^2$, where $x_{M,i} = (x_{i,j})_{j \in M}$ and $\hat{\beta}_M$ is the LS estimate of β_M obtained by fitting the model $y_i = x_{M,i}'\beta_M + \epsilon_i$, $i = 1, \ldots, 100$. How much time (in seconds) does it take to evaluate all the $Q_1(M)$'s, via the fast algorithm, for models M with dimension 10? and how much time (in seconds) does it take to do the same thing for $Q_2(M)$ (for which the fast algorithm does not apply)?

c.   Try a few more other dimensions (e.g., 5, 15, 20) and summarize the comparisons via a table.

5.2. This exercise is related to Example 5.1. Suppose that, instead, the actual gene-set scores are $X_1, \ldots, X_{m-1}, X_m + a$, where the X_i's are independent and distributed as $N(0, \sigma^2)$, where $\sigma^2 > 0$. Show that the probability that the last gene-set score is larger than any other gene-set scores decreases as m increases.

5.3. Referring to (ii) of the limited bootstrap procedure in Section 5.3. Explain why the smallest dimension K that allows "a peak in the middle" in the IF relative frequency plot is $K = 3$.

5.4. Carry out a simulation study to confirm the peak-pattern predicted in Example 5.3. Namely, simulate data in a way similar to simulation study presented in Section 5.3 so that there is one dominant gene set, three equally significant but non-dominant gene sets, and the rest of the gene sets are non-significant (zero). You may not need to perform the test for no gene set–simply compare the relative frequency plots with and without the dominant gene set. You may play around with the degree of dominance (compared to the non-dominant gene sets) to see how the patterns change.

5.5. Show that the covariance matrix of ϵ_i, defined below (5.4), has the expression (5.5).

5.6. Verify (5.11) for the case $K = 2$ and $m = 3$.

5.7. Verify expression (5.12). Also verify that the optimal mean function selected by the IF (near the end of Section 5.6) corresponds to (5.13).

Chapter 6

Fence Methods for Small Area Estimation and Related Topics

Small area estimation (SAE) has received increasing attention in recent literature. Here the term small area typically refers to a population for which reliable statistics of interest cannot be produced due to certain limitations of the available data. Examples of small areas include a geographical region (e.g., a state, county, municipality, etc.), a demographic group (e.g., a specific age × sex × race group), a demographic group within a geographic region, etc. In absence of adequate direct samples from the small areas, methods have been developed in order to "borrow strength". See Rao (2003) for a comprehensive account of various methods used in SAE. Statistical models, especially mixed effects models, have played important roles in SAE. Also see Jiang and Lahiri (2006) for an overview of mixed effects models in SAE.

While there is extensive literature on inference about small areas using mixed effects models, including estimation of small area means which is a problem of mixed model prediction, estimation of the mean squared error (MSE) of the empirical best linear unbiased predictor [EBLUP; see Rao (2003)], and prediction intervals [e.g., Chatterjee *et al.* (2007)], model selection in SAE has received much less attention. However, the importance of model selection in SAE has been noted by prominent researchers in this field [e.g., Battese *et al.* (1988), Ghosh and Rao (1994)]. Datta and Lahiri (2001) discussed a model selection method based on computation of the frequentist's Bayes factor in choosing between a fixed effects model and a random effects model. They focused on the following one-way balanced random effects model for the sake of simplicity: $y_{ij} = \mu + u_i + e_{ij}$, $i = 1, \ldots, m$, $j = 1, \ldots, k$, where the u_i's and e_{ij}'s are normally distributed with mean zero and variances σ_u^2 and σ_e^2, respectively. As noted by the authors, the choice between a fixed effects model and a random effects one in this case is equivalent to testing the following one-sided hypothesis H_0: $\sigma_u^2 = 0$ vs H_1: $\sigma_u^2 > 0$. The idea is further developed in Datta *et al.* (2011). Note that, however, not all model selection problems can be formulated as hypothesis testing problems. Fabrizi and Lahiri (2004) developed a robust model selection method in the context of complex surveys. Meza and Lahiri (2005) demonstrated the limitations of Mallows' C_p statistic in selecting the fixed covariates in a NER model [Battese *et al.* (1988)]. Simulation studies carried out by Meza and Lahiri (2005) showed that the C_p method without modification does not work well in the current mixed model setting when the variance σ_u^2 is large; on the other hand, a modified C_p criterion developed by these latter authors by adjusting the intra-cluster correlations performs similarly as the C_p in regression settings. It should be pointed out that all these studies are limited to linear mixed models, while model selection in SAE in a generalized linear mixed model (GLMM) setting has never been seriously addressed.

The fence methods provide a natural frame work for selecting an appropriate model for SAE. Note that a potential advantage of the fence is its flexibility in incorporating practical interest selecting the optimal model. The main interest of SAE is usually estimation, or prediction, of small area means. Consider, for example, the Fay-Herriot model [Fay and Herriot (1979); see Example 3.1]. In this case, the small area mean is the mixed effect, $\theta_i = x_i'\beta + v_i$, for the ith small area. Thus, the problem is that of prediction of a mixed effect, for which the standard method is empirical best linear unbiased prediction [EBLUP; e.g., Jiang (2007), sec. 2.3.1]. Suppose that the interest is to select the fixed covariates, which correspond to the components of x_i, among a number of candidates. One may treat this as a problem of mixed model selection, and apply the AF method. A simulation study in Example 3.1 showed that the AF has an outstanding performance in this case. Another class of mixed effects models that are extensively used in SAE is the NER model [Battese *et al.* (1988)]. Below we consider a case study under this model.

6.1   The NER model: A case study

Battese *et al.* (1988) presented data from 12 Iowa counties obtained from the 1978 June Enumerative Survey of the U.S. Department of Agriculture as well as data obtained from land observatory satellites on crop areas involving corn and soybeans. The objective was to predict the mean hectares of corn and soybeans per segment for the 12 counties using the satellite information. Their model can be expressed as (3.4), in which $x'_{ij}\beta = \beta_0 + \beta_1 x_{ij1} + \beta_2 x_{ij2}$; i represents county and j segment within the county; y_{ij} is the number of hectares of corn (or soybeans); x_{ij1} and x_{ij2} are the number of pixels classified as corn and soybeans, respectively, according to the satellite data. The mean hectares of crops can be expressed as $\theta_i = \bar{X}'_i\beta + v_i$, where $\bar{X}_{ij1}$ and $\bar{X}_{ij2}$ are the mean numbers of pixels classified as corn and soybeans per segment, respectively, which are available.

While fitting the NER model, Battese *et al.* (1988) (BHF) discussed possibility of including quadratic terms in $x'_{ij}\beta$. This raised an issue about variable selection, and motivates an application of the fence method. Following the latter authors' concern, the candidate variables are x_{ijk}, $k = 0, 1, \ldots, 5$, where $x_{ij0} = 1$; $x_{ij3} = x^2_{ij1}$, $x_{ij4} = x^2_{ij2}$, and $x_{ij5} = x_{ij1}x_{ij2}$. A simulation study under this setting has been carried out (see Example 3.3). Here we apply the AF after standardizing the data. The optimal models selected are, for the corn data, (3.4) with $x'_{ij}\beta = \beta_0 + \beta_1 x_{ij1}$; and, for the soybeans data, (3.4) with $x'_{ij}\beta = \beta_0 + \beta_2 x_{2ij}$. Comparing with the BHF models (and notice that the coefficients are, of course, different even though the same notations β are used), the similarity is that all models are linear in the satellite pixels (i.e., x_{ij1} and x_{ij2}). In other words, the quadratic terms are found unnecessary. As for the difference, we may summarize, in short, as follows. The BHF models: (1) corn by corn and soybeans; and (2) soybeans by corn and soybeans. Optimal models selected by the AF: (1) corn by corn; and (2) soybeans by soybeans. The models discovered by the AF are, at least, simple and intuitive. Figure 6.1 shows some interesting observations on how the models are selected, where the left plot corresponds to the corn, and right one to the soybeans. Notice the highest peaks in the middle of the plots. These correspond to the model selected by the adaptive fence method in each case. On each plot there is also a small peak (to the left of the highest peak in the middle), although the one for the soybeans is hardly visible. Interestingly, these small peaks correspond to the models of Battese *et al.* (1988) for the corn and soybeans, respectively. The fact that the smaller peak is less visible for the soybeans than for the corn can also be seen from the results of model fitting by Battese *et al.* (1988) Under their corn model, the estimated coefficients (standard errors) for the corn and soybeans pixels are .329 (.050) and $-.134$ (.056), respectively; under their soybeans model, the estimated coefficients (standard errors) for the corn and soybeans pixels are .028 (.058) and .494 (.065), respectively.

So far the model selection for both the Fay-Herriot model, discussed

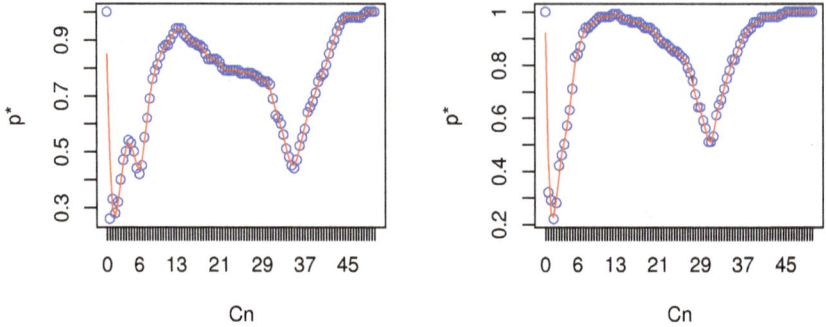

Fig. 6.1 *Left plot: p^* against $c = c_n$ for selecting the corn model. Right plot: p^* against $c = c_n$ for selecting the soybeans model.*

earlier in this chapter, and the NER model has not taken into account the special interest in SAE, as noted earlier. In the remaining sections of this chapter, this special interest will be considered.

6.2 Non-parametric model selection

Non-parametric small area models have received much attention in recent literature. In particular, Opsomer *et al.* (2011) proposed a spline-based non-parametric model for SAE. The idea is to approximate an unknown nonparametric small-area mean function by a penalized spline (P-spline). The authors then used a connection between P-splines and linear mixed models [Ruppert *et al.* (2003)] to formulate the approximating model as a linear mixed model, where the coefficients of the splines are treated as random effects. Consider, for simplicity, the case of univariate covariate. One may extend the Fay-Herriot model to the following

$$y_i = f(x_i) + b_i u_i + e_i, \quad i = 1, \ldots, m, \tag{6.1}$$

where y_i is the observation from the ith small area; x_i is a covariate; $f(\cdot)$ is an unknown (smooth) function; b_i is a known constant; u_i is a area-specific random effect, and e_i is a sampling error. It is assumed that u_i, e_i, $i = 1, \ldots, m$ are independent with $u_i \sim N(0, A)$, A being an unknown variance, and $e_i \sim N(0, D_i)$, D_i being a known sampling variance. Note that, besides $f(x_i)$, the model is the same as (a special case of) the Fay-

Herriot model. A P-spline approximation to f can be expressed as

$$\tilde{f}(x) = \beta_0 + \beta_1 x + \cdots + \beta_p x^p +$$
$$\gamma_1 (x - \kappa_1)_+^p + \cdots + \gamma_q (x - \kappa_q)_+^p, \qquad (6.2)$$

where p is the degree of the spline, q is the number of knots, κ_j, $1 \leq j \leq q$ are the knots, and $x_+ = x 1_{(x>0)}$. The coefficients β's and γ's are estimated by the penalized least squares, that is, by

$$\text{minimizing} \quad |y - X\beta - Z\gamma|^2 + \lambda |\gamma|^2, \qquad (6.3)$$

where $y = (y_i)_{1 \leq i \leq m}$, the ith row of X is $(1, x_i, x_i^p)$, the ith row of Z is $[(x_i - \kappa_1)_+^p, \ldots, (x_i - \kappa_q)_+^p]$, $i = 1, \ldots, m$, $j = 1, \ldots, n_i$, and λ is a penalty, or smoothing, parameter. To determine λ, Ruppert *et al.* (2003) used the following interesting connection a linear mixed model.

For simplicity of illustration, assume that $b_i = 0$, and that the e_i's are independent and distributed as $N(0, \tau^2)$. If the γ's are treated as random effects which are independent and distributed as $N(0, \sigma^2)$, the solution to (6.3) is the same as the best linear unbiased estimator (BLUE) for β, and the best linear unbiased predictor (BLUP) for γ, if λ is identical to the ratio τ^2/σ^2 (Exercise 6.1). Thus, the value of λ may be estimated by the ML, or REML, estimators of σ^2 and τ^2 [e.g., Jiang (2007)]. However, there has been study suggesting that this approach is biased towards undersmoothing (Kauermann 2005). Consider, for example, a special case in which $f(x)$ is, in fact, the quadratic spline with two knots given by

$$f(x) = 1 - x + x^2 - 2(x-1)_+^2 + 2(x-2)_+^2, \quad 0 \leq x \leq 3 \qquad (6.4)$$

(the shape is half circle between 0 and 1 facing up, half circle between 1 and 2 facing down, and half circle between 2 and 3 facing up). Note that this function is smooth in that it has a continuous derivative (Exercise 6.2). It is clear that, in this case, the best approximating spline should be $f(x)$ itself with only two knots, i.e., $q = 2$ (of course, one could use a spline with many knots to "approximate" the two-knot quadratic spline, but that would seem very inefficient in this case). However, if one uses the above linear mixed model connection, the ML (or REML) estimator of σ^2 is consistent only if $q \to \infty$ (i.e., the number of appearances of the spline random effects goes to infinity). The seemingly inconsistency has two worrisome consequences: (i) the meaning of λ may be conceptually difficult to interpret; (ii) the behavior of the estimator of λ may be unpredictable.

Clearly, a P-spline is characterized by p, q, and also the location of the knots. Note that, however, given p, q, the location of the knots can be selected by the space-filling algorithm implemented in R [**cover.design()**]. But the question how to choose p and q remains. The general "rule-of-thumb" is that p is typically between 1 and 3, and q proportional to the sample size, n, with 4 or 5 observations per knot. But there may still be a lot of choices given the rule-of-thumb. For example, if $n = 200$, the possible choices for q range from 40 to 50, which, combined with the range of 1 to 3 for p, gives a total of 33 choices for the P-spline.

The fence method offers a natural approach to choosing the degree of the spline, p, the number of knots, q, and the smoothing parameter, λ at the same time. Note, however, a major difference from the situations considered in Jiang *et al.* (2008) and Jiang *et al.* (2009) in that the true underlying model is not among the class of candidate models, i.e., the approximating splines (6.2). Furthermore, the role of λ in the model should be made clear: λ controls the degree of smoothness of the underlying model. A natural measure of lack-of-fit is $Q(M, \theta_M; y) = |y - X\beta - Z\gamma|^2$. However, $Q(M)$ is not obtained by minimizing $Q(M, \theta_M; y)$ over β and γ without constraint. Instead, we have $Q(M) = |y - X\hat{\beta} - Z\hat{\gamma}|^2$, where $\hat{\beta}$ and $\hat{\gamma}$ are the solution to (6.3), and hence depends on λ.

Another difference is that there may not be a full model among the candidate models. Therefore, the fence inequality (3.2) is replaced by

$$Q(M) - Q(\tilde{M}) \leq c, \tag{6.5}$$

where $\tilde{M}$ is the candidate model that has the minimum $Q(M)$. We use the following criterion of optimality within the fence which combines model simplicity and smoothness. For the models within the fence, choose the one with the smallest q; if there are more than one such models, choose the model with the smallest p. This gives the best choice of p and q. Once p, q are chosen, we choose the model within the fence with the largest λ. The tuning constant c is chosen adaptively using the AF idea (see Section 3.2), where parametric bootstrap is used for computing p^*. More specifically, $\tilde{M}$ and ML estimators under $\tilde{M}$ are used for the bootstrapping.

The following theorem is proved in Jiang (2010). For simplicity, assume that the matrix $W = (X \; Z)$ is of full rank. Let $P_{W^\perp} = I_n - P_W$, where $n = \sum_{i=1}^{m} n_i$ and $P_W = W(W'W)^{-1}W'$.

Theorem 6.1. Computationally, the above fence procedure is equivalent to the following: (i) first use the AF to select p and q using (6.5) with $\lambda = 0$ and $Q(M) = y'P_{W^\perp}y$ (see Lemma 6.1 below), and same criterion as above for choosing p, q within the fence; (ii) let M_0^* denotes the model corresponding to the selected p and q, find the maximum λ such that

$$Q(M_0^*, \lambda) - Q(\tilde{M}) \leq c^*, \tag{6.6}$$

where for any model M with the corresponding X and Z, we have

$$Q(M, \lambda) = |y - X\hat{\beta}_\lambda - Z\hat{\gamma}_\lambda|^2,$$
$$\hat{\beta}_\lambda = (X'V_\lambda^{-1}X)^{-1}X'V_\lambda^{-1}y,$$
$$\hat{\gamma}_\lambda = \lambda^{-1}(I_q + \lambda^{-1}Z'Z)^{-1}Z'(y - X\hat{\beta}_\lambda),$$
$$X'V_\lambda^{-1}X = X'X - \lambda^{-1}X'Z(I_q + \lambda^{-1}Z'Z)^{-1}Z'X,$$
$$X'V_\lambda^{-1}y = X'y - \lambda^{-1}X'Z(I_q + \lambda^{-1}Z'Z)^{-1}Z'y,$$

and c^* is chosen by the AF (V_λ is defined below but not directly needed here for the computation because of the last two equations).

Note that in step (i) of the Theorem one does not need to deal with λ. The motivation for (6.6) is that this inequality is satisfied when $\lambda = 0$, so one would like to see how far λ can go. In fact, the maximum λ is a solution to the equation $Q(M_0^*, \lambda) - Q(\tilde{M}) = c^*$. The purpose of the last two equations is to avoid direct inversion of $V_\lambda = I_n + \lambda^{-1}ZZ'$, whose dimension is equal to n, the total sample size. Note that V_λ does not have a block diagonal structure because of ZZ', so if n is large direct inversion of V_λ may be computationally burdensome.

The proof of the Theorem requires the following lemma, whose proof is left as an exercise (Exercise 6.3).

Lemma 6.1. For any M and y, $Q(M, \lambda)$ is an increasing function of λ with $\inf_{\lambda > 0} Q(M, \lambda) = Q(M)$.

The rest of the section is devoted to a simulation study designed to evaluate performance of the proposed fence method. We consider a simple case of (6.1) with $b_i = 1$. Furthermore, we assume $D_i = D$, $1 \leq i \leq m$. Then, the model can be expressed as

$$y_i = f(x_i) + \epsilon_i, \quad i = 1, \ldots, m, \tag{6.7}$$

where $\epsilon_i \sim N(0, \sigma^2)$ with $\sigma^2 = A + D$, which is unknown. Thus, the model is the same as a nonparametric regression model.

We consider three different cases that cover various situations and aspects. In the first case, Case 1, the true underlying function is a linear function, $f(x) = 1 - x$, $0 \leq x \leq 1$, hence the model reduces to the traditional Fay-Herriot model. The goal is to find out if fence can validate the traditional Fay-Herriot model in the case that it is valid. In the second case, Case 2, the true underlying function is the quadratic spline with two knots, given by (6.4). Here we intend to investigate whether the fence can identify the true underlying function in the "perfect" situation, that is, when $f(x)$ itself is a spline. The last case, Case 3, is perhaps the most

Table 6.1 **Nonparametric Model Selection - Case 1 & Case 2.** *Reported are empirical probabilities, in terms of percentage, based on 100 simulations that the optimal model is selected.*

Sample size	Case 1			Case 2		
	$m = 10$	$m = 15$	$m = 20$	$m = 30$	$m = 40$	$m = 50$
Highest Peak	62	91	97	71	83	97
Confidence L.B.	73	90	97	73	80	96

practical situation, in which no spline can provide a perfect approximation to $f(x)$. In other words, the true underlying function is not among the candidates. In this case $f(x)$ is chosen as $0.5 \sin(2\pi x)$, $0 \leq x \leq 1$, which is one of the functions considered by Kauermann (2005).

We consider situations of small or median sample size, namely, $m = 10$, 15 or 20 for Case I, $m = 30$, 40 or 50 for Case 2, and $m = 10$, 30 or 50 for Case 3. The covariate x_i are generated from the Uniform[0, 1] distribution in Case 1, and from Uniform[0, 3] in Case 2; then fixed throughout the simulations. Following Kauermann (2005), we let x_i be equidistant in Case 3. The error standard deviation σ in (6.7) is chosen as 0.2 in Case 1 and Case 2. This value is chosen such that the signal standard deviation in each case is about the same as the error standard deviation. As for Case 3, we consider three different values for σ, 0.2, 0.5 and 1.0. These values are also of the same order as the signal standard deviation in this case.

The candidate approximating splines for Case 1 and Case 2 are the following: $p = 0, 1, 2, 3$, $q = 0$ and $p = 1, 2, 3$, $q = 2, 5$ (so there are a total of 10 candidates). As for Case 3, following Kauermann (2005), we consider only linear splines (i.e., $p = 1$); furthermore, we consider the number of knots in the range of the "rule-of-thumb" (i.e., roughly 4 or 5 observations per knot; see section 1), plus the intercept model ($p = q = 0$) and the linear model ($p = 1$, $q = 0$). Thus, for $m = 10$, $q = 0, 2, 3$; for $m = 30$, $q = 0, 6, 7, 8$; and for $m = 50$, $q = 0, 10, 11, 12, 13$.

Table 5.1 shows the results based on 100 simulations under Case 1 and Case 2. As in Jiang *et al.* (2009), we consider both the highest peak, that is, choosing c with the highest p^*, and 95% lower bound, that is, choosing a smaller c corresponding to a peak of p^* in order to be conservative, if the corresponding p^* is greater than the 95% lower bound of the p^* for any larger c that corresponds to a peak of p^*. It is seen that performance of the AF is satisfactory even with the small sample size. Also, it appears that the confidence lower bound method works better in smaller sample, but makes almost no difference in larger sample. These are consistent with the findings of Jiang *et al.* (2009).

Table 6.2 **Nonparametric Model Selection - Case 3.** *Reported are empirical distributions, in terms of percentage, of the selected models.*

	Sample Size # of Knots	m = 10 0,2,3		m = 30 0,6,7,8		m = 50 0,10,11,12,13	
		(p,q)	%	(p,q)	%	(p,q)	%
σ = .2	Highest Peak	(0,0)	1	(1,0)	9	(1,10)	100
		(1,0)	31	(1,6)	91		
		(1,2)	68				
	Confidence L.B.	(1,0)	24	(1,0)	9	(1,10)	100
		(1,2)	76	(1,6)	91		
σ = .5	Highest Peak	(0,0)	14	(1,0)	21	(1,0)	13
		(1,0)	27	(1,6)	77	(1,10)	84
		(1,2)	56	(1,7)	2	(1,11)	2
		(1,3)	3			(1,12)	1
	Confidence L.B.	(0,0)	8	(1,0)	8	(1,0)	2
		(1,0)	23	(1,6)	89	(1,10)	94
		(1,2)	65	(1,7)	3	(1,11)	2
		(1,3)	4			(1,12)	2
σ = 1	Highest Peak	(0,0)	27	(0,0)	15	(0,0)	10
		(1,0)	20	(1,0)	18	(1,0)	26
		(1,2)	49	(1,6)	63	(1,10)	60
		(1,3)	4	(1,7)	4	(1,11)	2
						(1,12)	2
	Confidence L.B.	(0,0)	20	(0,0)	1	(0,0)	2
		(1,0)	13	(1,0)	13	(1,0)	13
		(1,2)	59	(1,6)	82	(1,10)	80
		(1,3)	8	(1,7)	4	(1,11)	2
						(1,12)	3

Table 6.2 shows the results for Case 3. Note that, unlike Case 1 and Case 2, here there is no optimal model (an optimal model must be a true model, according to our definition, which does not exist). So, instead of giving the empirical probabilities of selecting the optimal model, we give the empirical distribution of the selected models in each case. It is apparent that, as σ increases, the distribution of the models selected becomes more spread out. A reverse pattern is observed as m increases. The confidence lower bound method appears to perform better in picking up a model with splines. Within the models with splines, fence seems to overwhelmingly prefer fewer knots than more knots.

Note that the fence procedure allows us to choose not only p and q but also λ. In each simulation we compute $\hat{\beta} = \hat{\beta}_\lambda$ and $\hat{\gamma} = \hat{\gamma}_\lambda$, given below (6.6), based on the λ chosen by the adaptive fence. The fitted values are calculated by (6.2) with β and γ replaced by $\hat{\beta}$ and $\hat{\gamma}$, respectively. We then average the fitted values over the 100 simulations. Figure 6.2 shows

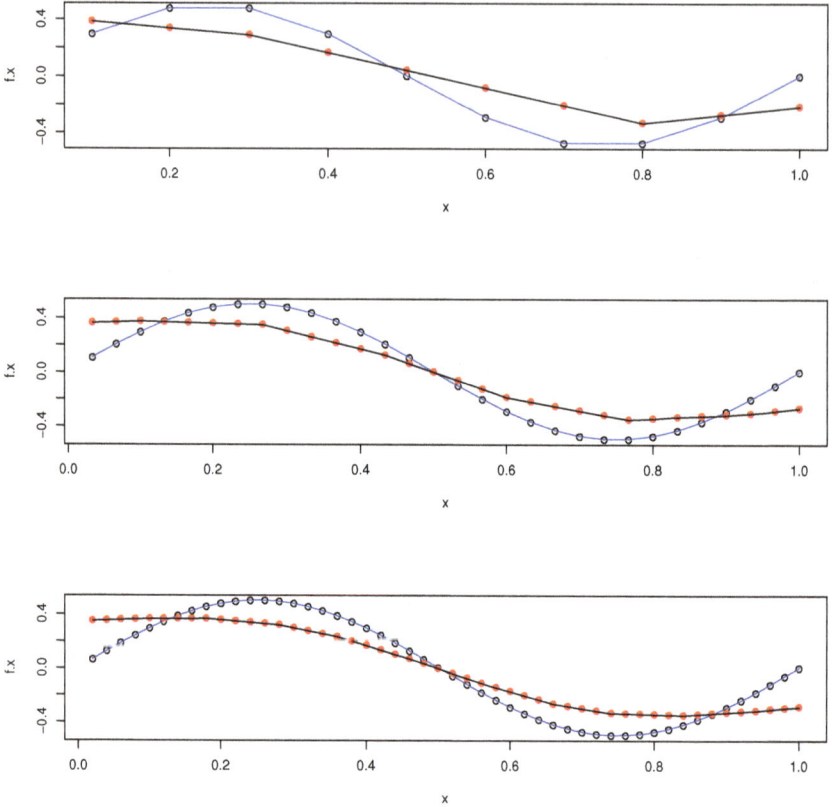

Fig. 6.2   *Case 3 Simulation. Top figure: Average fitted values for* $m = 10$. *Middle figure: Average fitted values for* $m = 30$. *Bottom figure: Average fitted values for* $m = 50$. *In all cases, the red dots represent the fitted values, while the blue circles correspond to the true underlying function.*

the average fitted values for the three cases ($m = 10, 30, 50$) with $\sigma = 0.2$ under Case 3. The true underlying function values, $f(x_i) = 0.5 \sin(2\pi x_i)$, $i = 1, \ldots, m$ are also plotted for comparison.

6.3   Another case study

We consider a dataset from Morris and Christiansen (1995) involving 23 hospitals (out of a total of 219 hospitals) that had at least 50 kidney transplants during a 27 month period (Table 6.3). The y_i's are graft failure rates

Table 6.3 **Hospital Data from Morris & Christiansen (1995).**

Area	y_i	x_i	$\sqrt{D_i}$
1	.302	.112	.055
2	.140	.206	.053
3	.203	.104	.052
4	.333	.168	.052
5	.347	.337	.047
6	.216	.169	.046
7	.156	.211	.046
8	.143	.195	.046
9	.220	.221	.044
10	.205	.077	.044
11	.209	.195	.042
12	.266	.185	.041
13	.240	.202	.041
14	.262	.108	.036
15	.144	.204	.036
16	.116	.072	.035
17	.201	.142	.033
18	.212	.136	.032
19	.189	.172	.031
20	.212	.202	.029
21	.166	.087	.029
22	.173	.177	.027
23	.165	.072	.025

for kidney transplant operations, that is, y_i = number of graft failures/n_i, where n_i is the number of kidney transplants at hospital i during the period of interest. The variance for graft failure rate, D_i, is approximated by $(0.2)(0.8)/n_i$, where 0.2 is the observed failure rate for all hospitals. Thus, D_i is assumed known. In addition, a severity index x_i is available for each hospital, which is the average fraction of females, blacks, children and extremely ill kidney recipients at hospital i. The severity index is considered as a covariate.

Ganesh (2009) proposed a Fay-Herriot model for the graft failure rates. as follows: $y_i = \beta_0 + \beta_1 x_i + v_i + e_i$, where the v_i's are hospital-specific random effects and e_i's are sampling errors. It is assumed that v_i, e_i are independent with $v_i \sim N(0, A)$ and $e_i \sim N(0, D_i)$. Here the variance A is unknown. Based on the model, the Ganesh obtained credible intervals for selected contrasts. However, inspections of the raw data uggest some nonlinear trends, which raises the question on whether the fixed effects part of the model can be made more flexible in its functional form.

To answer this question, we consider the Fay-Herriot model as a special member of a class of approximating spline models discussed in this section. More specifically, we assume the model (6.1) with $b_i = 1$ and x_i being the severity index. We then consider the approximating spline (6.2) with $p = 0, 1, 2, 3$ and $q = 0, 1, \ldots, 6$ ($p = 0$ is only for $q = 0$). Here the upper bound 6 is chosen according to the "rule-of-thumb" (because $m = 23$, so $m/4 = 5.75$). Note that the Fay-Herriot model corresponds to the case $p = 1$ and $q = 0$. The question is then to find the optimal model, in terms of p and q, from this class.

We apply the adaptive fence method described in Section 6.2 to this case. Here to obtain the bootstrap samples needed for obtaining c^*, we first compute the ML estimator under the model $\tilde{M}$, which minimizes $Q(M) = y' P_{W^\perp} y$ among the candidate models [i.e., (6.2); see Theorem 6.1], then draw parametric bootstrap samples under $\tilde{M}$ with the ML estimators treated as the true parameters. This is reasonable because $\tilde{M}$ is the best approximating model in terms of the fit, even though under model (6.1) there may not be a true model among the candidate models. The bootstrap sample size is chosen as 100.

The fence method has selected the model $p = 3$ and $q = 0$, that is, a cubic function with no knots, as the optimal model. We repeated the analysis 100 times, each time using different bootstrap samples. All results led to the same model: a cubic function with no knots. The left figure of Figure 6.3 shows the plot of p^* against $c = c_n$ in the AF model selection.

A few comparisons are always helpful. Our first comparison is to fence itself but with a more restricted space of candidate models. More specifically, we consider (6.2) with the restriction to linear splines only, i.e., $p = 1$, and knots in the range of the "rule-of-thumb", i.e., $q = 4, 5, 6$, plus the intercept model ($p = q = 0$) and the linear model ($p = 1$, $q = 0$). In this case, the fence method selected a linear spline with four knots (i.e., $p = 1$, $q = 4$) as the optimal model. The value of λ corresponding to this model is approximately equal to 0.001. The plot of p^* against c_n for this model selection is very similar to the left figure of Figure 6.3, and therefore omitted. In addition, the right figure of Figure 6.3 shows the fitted values and curves under the two models selected by the fence from within the different model spaces as well as the original data points.

A further comparison can be made by treating (6.1) as a generalized additive model [GAM; e.g., Hastie and Tibshirani (1990)] with heteroscedastic errors. A weighted fit can be obtained with the amount of smoothing optimized by using a generalized cross-validation (GCV) criterion. Here the

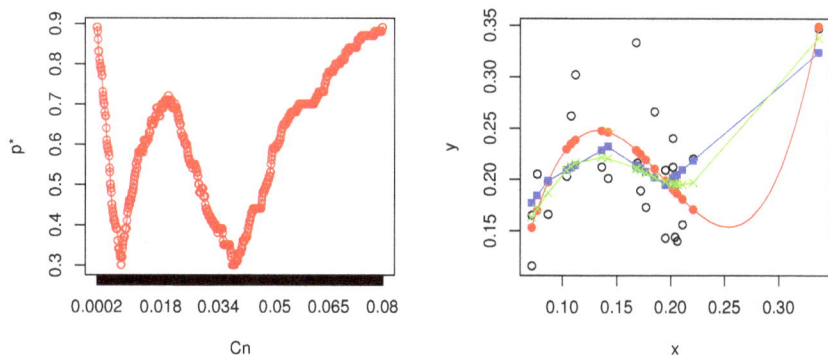

Fig. 6.3   Left: A plot of p^* against c_n from the search over the full model space. Right: The raw data and the fitted values and curves; red dots and curve correspond to the cubic function resulted from the full model search; blue squares and lines correspond to the linear spline with 4 knots resulted from the restricted model search; green X's and lines represent the GAM fits.

weights used are $w_i = 1/(A + D_i)$ where the ML estimate for A is used as a plug-in estimate. Recall that the D_i's are known. This fitted function is also overlayed in the right plot of Figure 6.3. Notice how closely this fitted function resembles the restricted space fence fit.

To expand the class of models under consideration by GCV-based smoothing, we used the BRUTO procedure [Hastie and Tibshirani (1990)] which augments the class of models to look at a null fit and a linear fit for the spline function; and embeds the resulting model selection (i.e., null, linear or smooth fits) into a weighted backfitting algorithm using GCV for computational efficiency. Interestingly here, BRUTO finds simply an overall linear fit for the fixed effects functional form. While certainly an interesting comparison, BRUTO's theoretical properties for models like (6.1) have not really been studied in depth.

Finally, as mentioned in Section 6.2, by using the connection between P-spline and linear mixed model one can formulate (6.2) as a linear mixed model, where the spline coefficients are treated as random effects. The problem then becomes a (parametric) mixed model selection problem, hence the method of Section 3.2 can be applied. In fact, this was our initial approach to this dataset, and the model we found was the same as the one by BRUTO. However, we have some reservation about this approach, as explained in Section 6.2.

6.4 Predictive model selection

In many cases, the problem of interest is prediction rather than estimation, and one of those cases is SAE, as mentioned. More generally, problems of practical interest often arise in the context of mixed model prediction, in which the quantity of interest can be expressed in terms of a mixed effect. See, for example, Robinson (1991), Jiang and Lahiri (2006), for reviews on prediction of mixed effects and its applications. As noted, the fence is flexible in choosing the criterion of optimality in selecting the model within the fence to incorporate consideration of practical interest. Another potentially useful feature of the fence, which so far has not been seriously explored, is that the fence is also flexible in choosing the measure of lack-of-fit, Q, in (3.2), to incorporate the practical interest. Because prediction is currently of main interest, it makes sense to use a predictive measure, rather than an estimation-based measure. Examples of the latter include the negative log-likelihood (Example 1.3) and the RSS (Example 1.5).

We now describe a general approach to deriving a predictive measure of lack-of-fit by considering two special cases.

Let us first consider a general problem of linear mixed model prediction [e.g., Robinson (1991)]. The assumed model is

$$y = X\beta + Zv + e, \tag{6.8}$$

where X, Z are known matrices; β is a vector of fixed effects; a vector of fixed effects; v, e are vectors of random effects and errors, respectively, such that $v \sim N(0, G)$, $e \sim N(0, \Sigma)$, and v, e are uncorrelated. An important issue for model-based statistical inference is the possibility of model misspecificaion. To take the latter into account, suppose that the true underlying model is

$$y = \mu + Zv + e, \tag{6.9}$$

where $\mu = \mathrm{E}(\mathbf{y})$. Here, E represents expectation with respect to the true distribution of y, which may be unknown but is not model-dependent. So, if $\mu = X\beta$ for some β, the model is correctly specified; otherwise, the model is misspecified. Our interest is prediction of a vector of mixed effects that can be expressed as

$$\theta = F'\mu + R'v, \tag{6.10}$$

where F, R are known matrices. We consider some examples.

Example 6.1. Robinson (1991) used the following example to illustrate the BLUP methd. A linear mixed model is assumed for the first lactation yields of dairy cows with sire additive genetic merits being treated as random effects and herd effects as fixed effects. The herd effects are represented by $\beta_i, i = 1, 2, 3$, and the sire effects by $v_j, j = 1, 2, 3, 4$. The assumed model can be expressed as

$$y_{ijk} = \beta_i + v_j + e_{ijk}, \quad (i,j) \in S, \; k = 1, \ldots, n_{ij}, \tag{6.11}$$

where $S = \{(1,1), (1,4), (2,2), (2,4), (3,3), (3,4)\}$, $n_{11} = n_{14} = n_{22} = 1$, and $n_{24} = n_{33} = n_{34} = 2$. Here the different k's for the same (i,j) correspond to different daughters of the same sire within the same herd. Furthermore, e_{ij} represent the environmental errors. It is assumed that the sire effects and environmental errors are independent such that $v_j \sim N(0, \sigma^2)$ and $e_{ij} \sim N(0, \tau^2)$. The assumed model can be written as (6.8) with $\beta = (\beta_1, \beta_2, \beta_3, \beta_4)'$, $v = (v_1, v_2, v_3, v_4)'$, $e = (e_{111}, e_{141}, e_{221}, e_{241}, e_{242}, e_{331}, e_{332}, e_{341}, e_{342})'$,

$$X = \begin{pmatrix} 1 & 0 & 0 \\ 1 & 0 & 0 \\ 0 & 1 & 0 \\ 0 & 1 & 0 \\ 0 & 1 & 0 \\ 0 & 0 & 1 \\ 0 & 0 & 1 \\ 0 & 0 & 1 \\ 0 & 0 & 1 \end{pmatrix} \quad \text{and} \quad Z = \begin{pmatrix} 1 & 0 & 0 & 0 \\ 0 & 0 & 0 & 1 \\ 0 & 1 & 0 & 0 \\ 0 & 0 & 0 & 1 \\ 0 & 0 & 0 & 1 \\ 0 & 0 & 1 & 0 \\ 0 & 0 & 1 & 0 \\ 0 & 0 & 0 & 1 \\ 0 & 0 & 0 & 1 \end{pmatrix}.$$

Furthermore, we have $G = \sigma^2 I_4$ and $\Sigma = \tau^2 I_9$, where I_n denotes the n-dimensional identity matrix (Exercise 6.4). Suppose that there is interest in predicting the genetic merits of the sires. Thus, the quantity of interest is v, which can be expressed as (6.10) with $\theta = v$, $F = 0$ and $R = I_4$. If, instead, one is interested in the sire-specific mean yields, averaged over different herds, let $\mu_{ijk} = \mathrm{E}(y_{ijk})$. Then, the quantities of interest may be expressed as $\theta_1 = \mu_{111} + v_1$, $\theta_2 = \mu_{221} + v_2$, $\theta_3 = (\mu_{331} + \mu_{332})/2 + v_3$, and $\theta_4 = (\mu_{141} + \mu_{241} + \mu_{242} + \mu_{341} + \mu_{342})/5 + v_4$, or (6.10) with $\theta = (\theta_1, \theta_2, \theta_3, \theta_4)'$,

$$F = \begin{pmatrix} 1 & 0 & 0 & 0 & 0 & 0 & 0 & 0 & 0 \\ 0 & 0 & 1 & 0 & 0 & 0 & 0 & 0 & 0 \\ 0 & 0 & 0 & 0 & 0 & 1/2 & 1/2 & 0 & 0 \\ 0 & 1/5 & 0 & 1/5 & 1/5 & 0 & 0 & 1/5 & 1/5 \end{pmatrix}'$$

and $R = I_4$, where μ is defined similarly as the e vector.

Example 6.2. Recall that the Fay-Herriot model [Fay and Herriot (1979)] was introduced to estimate the per-capita income of small places with population size less than 1,000:

$$y_i = x_i'\beta + v_i + e_i, \quad i = 1, \ldots, m,$$

where x_i is a vector of known covariates, β is a vector of unknown regression coefficients, v_i's are area-specific random effects and e_i's are sampling errors. It is assumed that v_i's, e_i's are independent with $v_i \sim N(0, A)$ and $e_i \sim N(0, D_i)$. The variance A is unknown, but the sampling variances D_i's are assumed known. The assumed model can be expressed as (6.8) with $X = (x_i')_{1 \leq i \leq m}$, $Z = I_m$, $G = AI_m$ and $\Sigma = \text{diag}(D_1, \ldots, D_m)$. The problem of interest is estimation of the small area means. Let $\mu_i = \text{E}(y_i)$. Then, the small area means can be expressed as $\theta_i = \text{E}(y_i|v_i) = \mu_i + v_i$, under the true underlying model (6.9). Here, again, E represents the true conditional expectation rather than conditional expectation under the assumed model. See Jiang *et al.* (2011a) for more details. Thus, the quantity of interest can be expressed as (6.10) with $\theta = (\theta_i)_{1 \leq i \leq m}$, and $F = R = I_m$.

For simplicity, assume that both G and Σ are known. Then, under the assumed model, the best predictor (BP) of θ, in the sense of minimum mean squared prediction error (MSPE), is the conditional expectation,

$$\begin{aligned} \text{E}_M(\theta|y) &= F'\mu + R'\text{E}_M(v|y) \\ &= F'X\beta + R'GZ'V^{-1}(y - X\beta), \end{aligned} \quad (6.12)$$

where $V = \Sigma + ZGZ'$ and β is the true vector of fixed effects [e.g., Jiang (2007), p. 75]. The E_M in (6.12) denotes conditional expectation under the assumed model, (6.8), rather than the true model (6.9). Although model (6.8) may be subject to model misspecification, it is usually (much) simpler and utilizes the available covariates, X. On the other hand, even if model (6.9) is correct, or close to be correct, it is too broad to be useful; furthermore, it does not make use of any of the available covariates, which is often practically unacceptable. For these reasons, the assumed model, (6.8), is always the one of main interest. In other words, one cannot abandon the assumed model; all one could do is to try to do the best under the assumed model, that is, to find the best way to estimate the parameters, in this case β, under the assumed model. It should be noted that, for now, we put aside the issue about model selection, for the assumed model, which we shall discuss in the sequel. The question then is: What is the role that the true model, (6.9), plays in this business? The answer is that the true model can help to determine the best way to estimate the parameters so that it is more robust to model misspecification. More specifically, the true model is used in evaluating the predictive performance of the BP, (6.12), to make sure that the evaluation is fair and not model-dependent. This is why, intuitively, the resulting estimator of β is more robust to model misspecification Jiang *et al.* (2011a). The idea introduced here is particularly important when dealing with model selection, because, obviously, the measure of lack-of-fit has to be objective, or "fair", to all of the candidate models.

To derive the best estimator of β, write $B = R'GZ'V^{-1}$ and $\Gamma = F' - B$. Let $\tilde{\theta}$ denote the right side of (6.12), where β is understood as a parameter vector to be determined. The predictive performance of $\tilde{\theta}$ is typically measured by the MSPE, defined as $\text{MSPE}(\tilde{\theta}) = \text{E}(|\tilde{\theta} - \theta|^2)$. Here, again, E denotes expectation under the true model. It can be shown (Exercise 6.5) that the MSPE can be expressed, alternatively, as

$$\text{MSPE}(\tilde{\theta}) = \text{E}\{(y - X\beta)'\Gamma'\Gamma(y - X\beta) + \cdots\}, \tag{6.13}$$

where $\cdots$ does not depend on β. Here comes another key point: Unlike $\text{E}(|\tilde{\theta} - \theta|^2)$, the expression inside the expectation on the right side of (6.13) involves a function of the observed data and β (and nothing else), and something unrelated to β. Therefore, it is natural to estimate β by minimizing the expression inside the expectation on the right side of (6.13), which is equivalent to minimizing the expression without $\cdots$. This leads to what we call the best predictive estimator, or BPE, of β, given by

$$\hat{\beta}_{\text{BPE}} = (X'\Gamma'\Gamma X)^{-1}X'\Gamma'\Gamma y, \tag{6.14}$$

assuming that $\Gamma'\Gamma$ is nonsingular and X is full rank. As a comparison, the ML estimator (MLE) of β, under the assumed model, is given by

$$\hat{\beta}_{\text{MLE}} = (X'V^{-1}X)^{-1}X'V^{-1}y, \tag{6.15}$$

assuming nonsingularity of V. Again, we consider an example.

Example 6.2 (continued). Under the Fay-Herriot model, let $r_i = D_i/(A + D_i)$. Then, the expressions (6.14) and (6.15) simplified to

$$\hat{\beta}_{\text{BPE}} = \left(\sum_{i=1}^{m} r_i^2 x_i x_i'\right)^{-1} \sum_{i=1}^{m} r_i^2 x_i y_i, \tag{6.16}$$

$$\hat{\beta}_{\text{MLE}} = \left(\sum_{i=1}^{m} \frac{x_i x_i'}{A + D_i}\right)^{-1} \sum_{i=1}^{m} \frac{x_i y_i}{A + D_i}. \tag{6.17}$$

An interesting observation on the latest expressions is that both the BPE and the MLE are weighted averages of the data; however, the BPE gives more weights to data points with larger sampling variances; on the other hand, the MLE does just the opposite, by giving more weights to the data points with smaller sampling variances. The question is: Who is right? Recall the "business" here is to find the best estimator to substitute the β in the expression of BP. In this case, the BP, (6.12), can be expressed as

$$\mathrm{E}_M(\theta|y) = \frac{A}{A + D_i} y_i + \frac{D_i}{A + D_i} x_i' \beta.$$

It is seen that, in this expression, β is only involved in the second term, which carries the weight $D_i/(A + D_i)$. Thus, the larger the sampling variance, D_i, the more weight is given to the second term. In other words, as far as the BP is concerned, the estimation of β is more relevant to those areas i with larger D_i. Therefore, intuitively, the rationale of the BPE seems to make more sense. Imagine that a meeting is being held with the presence of representatives from the m small areas. The agenda of the meeting is to determine the weights in the β estimator, then substitute this estimator in the BP. The representatives from those areas with larger sampling variances argue that, because the BP is more relevant to their business, their "voice" should be heard more; therefore, more weights should be given to their areas. The argument appears to be reasonable.

When the β in the BP is substituted by the BPE, the result is called the observed best predictor, or OBP [Jiang *et al.* (2011a)]. The latter authors showed that the OBP generally outperforms the traditional empirical best linear unbiased predictor [EBLUP, e.g., Rao (2003)] when the underlying model is misspecified. As far as this section is concerned, our main interest is to develop a fence method that take into account the particular interest of mixed model prediction. The development so far have set up the stage for this mission. Namely, one can define the measure of lack-of-fit, $Q(M)$, as the minimizer, over β, of the expression without $\cdots$ inside the expectation on the right side of (6.13). Clearly, this measure is designed specifically for the mixed model prediction problem. Also, when it comes to model selection, it is important that the measure of lack-of-fit is "fair" to every candidate model. The above measure $Q(M)$ has this feature, because the expectation in (6.13) is under an objective true model. Once we have the measure Q, we can use it in (3.2) for the fence.

The assumption that G and Σ are known can be relaxed, to some extent, especially for G. In fact, it can be shown that essentially the same derivation of Jiang *et al.* (2011a) goes through, and the resulting measure of lack-of-fit, $Q(M)$, is the minimizer of $(y - X\beta)'\Gamma'\Gamma(y - X\beta) - 2\mathrm{tr}(\Gamma'\Sigma)$, assuming that Σ is known (Exercise 6.6). We consider some examples.

Example 6.2 (continued). For the Fay-Herriot model with A unknown, it can be shown that the right side of (6.13) is equal to

$$E\{(y - X\beta)'\Gamma^2(y - X\beta) + 2A\text{tr}(\Gamma) - \text{tr}(D)\}, \qquad (6.18)$$

where $D = \text{diag}(D_i, 1 \leq i \leq m)$ (Exercise 6.6). Note that, in this case, $\Gamma = \text{diag}(r_i, 1 \leq i \leq m)$. Thus, the measure of lack-of-fit, $Q(M)$, is the minimizer of the expression inside the expectation in (6.18) over β and A. The minimization can be done in two steps. First minimize the expression over β with A fixed, resulting $\tilde{Q}(A) = y'\Gamma P_{(\Gamma X)^\perp}\Gamma y + 2A\text{tr}(\Gamma)$, where for any matrix M, $P_{M^\perp} = I - P_M$ with $P_M = M(M'M)^{-1}M'$ (assuming nonsingularity of $M'M$), and I is the identity matrix. It follows that $P_{(\Gamma X)^\perp} = I - \Gamma X(X'\Gamma^2 X)^{-1}X'\Gamma$ (Exercise 6.6). The next step is to minimize $\tilde{Q}(A)$ with respect to $A \geq 0$. This is typically done numerically.

Example 6.3 (NER model). Recall the NER model of Section 6.1. The model may be expressed, in general, as

$$y_{ij} = x'_{ij}\beta + v_i + e_{ij}, \qquad (6.19)$$

$i = 1, \ldots, m, j = 1, \ldots, n_i$, where the v_i's are the area-specific random effects and e_{ij}'s are errors which are assumed to be independent and $\text{var}(e_{ij}) = \sigma_e^2$, where σ_v^2 and σ_e^2 are unknown. Under the NER model, the small area mean, assuming infinite population, is $\theta_i = \bar{X}'_i\beta + v_i$ for the ith small area, where $\bar{X}_i$ is the population mean of the x_{ij}'s [assumed known; e.g., Rao (2003)]. It is seen that θ_i is a (linear) mixed effect. Let $\gamma = \sigma_v^2/\sigma_e^2$. Then, BP of θ_i, in the sense of minimum MSPE, can be expressed as

$$\tilde{\theta}_i = \bar{X}'_i\beta + \frac{n_i\gamma}{1 + n_i\gamma}(\bar{y}_{i\cdot} - \bar{x}'_{i\cdot}\beta), \qquad (6.20)$$

where β and γ are the true parameters, $\bar{y}_{i\cdot} = n_i^{-1}\sum_{j=1}^{n_i} y_{ij}$ and $\bar{x}_{i\cdot} = n_i^{-1}\sum_{j=1}^{n_i} x_{ij}$ (Exercise 6.7). The traditional BLUP method is based (6.20) with β replaced by its MLE, assuming that γ is known; the EBLUP is derived from the BLUP with γ replaced by a consistent estimator, such as the ML or REML estimator [e.g., Jiang (2007)].

Note that, in practice, the small area populations are finite. Thus, following Jiang *et al.* (2011a), we consider a super-population NER model. Suppose that the subpopulations of responses $\{Y_{ik}, k = 1, \ldots, N_i\}$ and auxiliary data $\{X_{ikl}, k = 1, \ldots, N_i\}, l = 1, \ldots, p$ are realizations from corresponding super-populations that are assumed to satisfy the NER model:

$$Y_{ik} = X'_{ik}\beta + v_i + e_{ik}, \ i = 1, \ldots, m, k = 1, \ldots, N_i, \qquad (6.21)$$

where β, v_i and e_{ik} satisfy the same assumptions as in (6.19). Under the finite-population setting, the true small area is $\theta_i = \bar{Y}_i = N_i^{-1}\sum_{k=1}^{N_i} Y_{ik}$

(as opposed to $\theta_i = \bar{X}'_i\beta + v_i$ under the infinite-population setting) for $1 \leq i \leq m$. Furthermore, write $\rho_i = n_i/N_i$. Then, the finite-population version of the BP (6.20) is [e.g., Rao (2003), sec. 7.2.5]

$$\tilde{\theta}_i = \bar{X}'_i\beta + \left\{\rho_i + (1 - \rho_i)\frac{n_i\gamma}{1 + n_i\gamma}\right\}(\bar{y}_{i\cdot} - \bar{x}'_{i\cdot}\beta), \qquad (6.22)$$

where β and γ are the true parameters.

In Example 6.2, derivation of the measure Q is based on a model-based MSPE under the broader model (6.9). Here, we shall consider a different type of MSPE, called design-based MSPE. The latter means that the MSPE, or, more generally, expectations, are taken with respect to the randomness of sampling from the small areas. Here, for simplicity, simple random sampling is considered [e.g., Fuller (2009), sec. 1.2]. Note that the design-based MSPE is not available in the case of area-level data, such as Example 6.2, because the actual samples from the areas are not available (only summaries of the data, y_i, are available at the area level). This is why, in Example 6.2, the MSPE was model-based, but was evaluated under a broader model than the assumed. In general, a "rule of thumb" is that one should make the MSPE as model-free as possible, so that it would be objective and (relatively) robust to model-misspecifications. The design-based MSPE is defined by $\text{MSPE}(\tilde{\theta}) = \text{E}_d(|\tilde{\theta} - \theta|^2)$, where E_d means design-based expectation. It can be shown (Exercise 6.8) that we have

$$\text{MSPE}(\tilde{\theta}) = \text{E}_d\left[\sum_{i=1}^{m}\left\{\tilde{\theta}_i^2 - 2a_i(\sigma^2)\bar{X}'_i\beta\bar{y}_{i\cdot} + b_i(\sigma^2)\hat{\mu}_i^2\right\}\right], \qquad (6.23)$$

where $\sigma^2 = (\sigma_v^2, \sigma_e^2)'$, $a_i(\sigma^2) = (1 - \rho_i)\sigma_e^2/(\sigma_e^2 + n_i\sigma_v^2)$, $b_i(\sigma^2) = 1 - 2\{\rho_i + (n_i\sigma_v^2/\sigma_e^2)a_i(\sigma^2)\}$, and $\hat{\mu}_i^2$ is a design-unbiased estimator of μ_i^2 given by

$$\hat{\mu}_i^2 = \frac{1}{n_i}\sum_{j=1}^{n_i}y_{ij}^2 - \frac{N_i - 1}{N_i(n_i - 1)}\sum_{j=1}^{n_i}(y_{ij} - \bar{y}_{i\cdot})^2. \qquad (6.24)$$

Here design-unbiasedness means that $\text{E}_d(\hat{\mu}_i^2) = \theta_i^2 = (\bar{Y}_i)^2$ (Exercise 6.8). This, the measure of lack-of-fit, $Q(M)$, is the minimizer of the expression inside the E_d in (6.23). A similar computational procedure to that described in Example 6.2 (continued) can be developed (Exercise 6.8).

Example 6.4 (count data). In many cases, the data for the response variables are counts [e.g., Münnich *et al.* (2009)]. A linear model for the responses, as considered above, may not be appropriate in such cases. For simplicity, suppose that the responses are counts, denoted by y_i, and that, in addition, a vector of covariates, x_i, is also available. The model of interest assumes that, given the random effects, v_i, y_i has a Poisson distribution with mean μ_i, such that

$$\log(\mu_i) = x_i'\beta + v_i. \tag{6.25}$$

The BP of μ_i, under the assumed model, can be expressed as

$$\mathrm{E}_{M,\psi}(\mu_i|y) = g_i(\psi, y_i), \tag{6.26}$$

where $\mathrm{E}_{M,\psi}$ denotes conditional expectation under the assumed model, M, and parameter vector, ψ, under M, and $g_i(\cdot,\cdot)$ is a known function which does not have an analytic expression. Nevertheless, the g function can be evaluated numerically fairly easily. Following a similar idea as in Example 6.2 (continued), we evaluate the performance of the BP under a broader model, which states that, conditioning on μ_i, y_i is Poisson(μ_i), but the expression (6.25) is not assumed. In other words, under the broader model, the μ_i's are completely unspecified. Write $\mu = (\mu_i)_{1\le i\le m}$, and $\tilde{\mu} = (\tilde{\mu}_i)_{1\le i\le m}$, where $\tilde{\mu}_i$ is the right side of (6.26) when ψ is considered as an unknown parameter vector. Consider

$$\begin{aligned}
\mathrm{MSPE} &= \mathrm{E}(|\tilde{\mu} - \mu|^2) \\
&= \sum_{i=1}^{m} \mathrm{E}\{g_i(\psi, y_i) - \mu_i\}^2 \\
&= \mathrm{E}\left\{\sum_{i=1}^{m} g_i^2(\psi, y_i)\right\} - 2\sum_{i=1}^{m} \mathrm{E}\{g_i(\psi, y_i)\mu_i\} + \sum_{i=1}^{m} \mathrm{E}(\mu_i^2) \\
&= I_1 - 2I_2 + I_3, \tag{6.27}
\end{aligned}$$

where E denotes expectation under the broader model. Note that I_3 does not involve ψ, even though it may be completely unknown. It can be shown (Exercise 6.9) that

$$\mathrm{E}\{g_i(\psi, y_i)\mu_i\} = \sum_{k=0}^{\infty} g_i(\psi, k)(k+1)\mathrm{E}\{1_{(y_i=k+1)}\}, \tag{6.28}$$

where 1_A is the indicator of even A ($= 1$ is A occurs, and 0 otherwise). Thus, if we define $g_i(\psi, -1) = 0$, we have

$$\begin{aligned}
\mathrm{E}\{g_i(\psi, y_i)\mu_i\} &= \mathrm{E}\left\{\sum_{k=0}^{\infty} g_i(\psi, k)(k+1)1_{(y_i=k+1)}\right\} \\
&= \mathrm{E}\{g_i(\psi, y_i - 1)y_i\}.
\end{aligned}$$

Therefore, combined with (6.27), we have

$$\mathrm{MSPE} = \mathrm{E}\left\{\sum_{i=1}^{m} g_i^2(\psi, y_i) - 2\sum_{i=1}^{m} g_i(\psi, y_i - 1)y_i + \cdots\right\}, \tag{6.29}$$

where $\cdots$ does not depend on ψ. A predictive measure of lack-of-fit, $Q(M)$, is obtained by minimizing the expression inside the expectation of (6.29), but without $+\cdots$, with respect to ψ.

We have shown how to obtain a predictive measure of lack-of-fit, Q, in various cases. We can use this Q in the fence inequality, e.g., (3.2), to select the model, if the cut-off, c, is given. For example, if minimum dimension (or parsimony) is the criterion of optimality for selecting the model within the fence, then, among the models within the fence, that is, those that satisfy (3.2), we choose the model that has the minimum dimension. This is what we call *predictive fence*. To apply the AF idea to the best interest of prediction, another modification seems plausible. Recall that the AF (see Section 3.2) selects the optimal cut-off, e.g., c in (3.2), by maximizing the empirical (bootstrap) probability of the most frequently selected model. As noted (see Section 3.1), this is similar to the maximum likelihood idea, but, here, the main concern is prediction accuracy. It seems that, instead of choosing c that minimizes (3.1), one should be choosing c that minimizes

$$\mathrm{mspe}(c) = \mathrm{E}(|\hat{\theta}_c - \theta|^2), \qquad (6.30)$$

where θ is the true vector of mixed effects, and $\hat{\theta}_c$ is the predictor of θ, obtained via a certain procedure, based on the model selected by the fence using the predictive measure, Q, and the given cut-off c. Similar to (3.1), here, again, two things are unknown. The first is E, that is, under what distribution should the expectation in (6.30) be evaluated? This problem may be dealt with using the bootstrap idea as for the AF. However, it is not clear, at this point, how to deal with the second unknown, that is, θ, the true mixed effect. Note that the latter is not model-dependent. The previous AF idea, motivated by the maximum likelihood, does not seem to apply here. The performance of predictive fence, especially regarding choosing the cut-off, c, remains largely unexplored at this point.

6.5   Exercises

6.1. This exercise requires some knowledge about the BLUE and BLUP in linear mixed model [e.g., Jiang (2007), sec. 2.3.1]. For simplicity of illustration, assume that $b_i = 0$ in (6.1), and that the e_i's are independent and distributed as $N(0, \tau^2)$. If the γ's are treated as random effects which are independent and distributed as $N(0, \sigma^2)$, the solution to (6.3) is the same as the best linear unbiased estimator (BLUE) for β, and the best linear unbiased predictor (BLUP) for γ, if λ is identical to the ratio τ^2/σ^2.

6.2. Consider the quadratic spline given by (6.4). Show that the shape of the spline is half circle between 0 and 1 facing up, half circle between 1 and 2 facing down, and half circle between 2 and 3 facing up. Also show that the function is smooth in that it has a continuous derivative.

6.3. Prove Lemma 6.1.

6.4. Verify that, in Example 6.1, the assumed model (6.11) is a special case of (6.8), with $G = \sigma^2 I_4$ and $\Sigma = \tau^2 I_9$.

6.5. Show that, in Section 6.4, the MSPE can be expressed as (6.13).

6.6. This exercise is related to some derivations in Section 6.4.

a. Show that, assuming that Σ is known, essentially the same derivation as (6.13) goes through [see Jiang *et al.* (2011a) for detail], resulting the measure of lack-of-fit, $Q(M)$, which is the minimizer of $(y - X\beta)'\Gamma'\Gamma(y - X\beta) - 2\mathrm{tr}(\Gamma'\Sigma)$.

b. In particular, verify (6.18) with $\Gamma = \mathrm{diag}(r_i, 1 \leq i \leq m)$ and $r_i = D_i/(A + D_i)$.

c. Show that, with A fixed, the minimizer of the expression inside (6.18) is $\tilde{Q}(A) = y'\Gamma P_{(\Gamma X)^\perp}\Gamma y + 2A\mathrm{tr}(\Gamma)$ [see the notation below (6.18)].

6.7. Show that, in Example 6.3, the BP of θ_i can be expressed as (6.20).

6.8. This exercise is related to Example 6.3.

a. Verify expression (6.23).

b. Show the design-unbiasedness of $\hat{\mu}_i^2$, that is, $\mathrm{E}_\mathrm{d}(\hat{\mu}_i^2) = \theta_i^2 = (\bar{Y}_i)^2$.

c. Develop a similar two-step computational procedure to that in Example 6.2 (continued), below (6.18).

6.9. Verify (6.28).

Chapter 7

Shrinkage Selection Methods

Simultaneous variable selection and estimation by penalized likelihood methods have received considerable interest in the recent literature on statistical learning [see Fan and Lv (2010) for an overview]. In a seminal paper, Tibshirani (1996) proposed the least absolute shrinkage and selection operator (Lasso) to automatically select variables via continuous shrinkage. The method has gained popularity thanks to the developments of several computation algorithms [see Fu (1998), Osborne *et al.* (2000), and Efron *et al.* (2004)]. In particular, the LARS algorithm proposed by Efron *et al.* (2004) finds the entire regularization path at the same computational cost of one single ordinary LS fit. Fan and Li (2001) proposed the smoothly clipped absolute deviation method (SCAD) which has the *oracle property* under certain conditions. For the most part, the oracle property means that the parameter estimators under the selected model are as efficient as those when the true underlying model is known. Local quadratic method [Fan and Li (2001)], minorize-maximize [MM; Hunter and Li (2005)], local linear approximation method (LLA; Zou and Li (2008)] have been proposed to compute the SCAD estimates. In particular, the LLA reduces the computational cost for the SCAD dramatically by applying the LARS algorithm. Furthermore, Zou (2006) proposed the adaptive Lasso and shows it also has the oracle property. In this chapter, we discussed a few problems for which solutions were obtained, or may be sought, using the fence idea. Although these methods may seem somewhat different from the fence, it may be argued, as we shall see, that they are, indeed, special cases of the fence.

7.1   Selection of regularization parameter

The selection of the regularization parameter, or tuning parameter, in the penalized likelihood approach is an important problem which determines the dimension of the selected model. Consider, for example, a linear regression model: $y = X\beta + \epsilon$, where $y = (y_1, \ldots, y_n)'$ is an $n \times 1$ vector of response, $X = (x_1, \ldots, x_p)$ is an $n \times p$ matrix of predictors, $\beta = (\beta_1, \ldots, \beta_p)'$ is the vector of regression coefficients, and ϵ is a vector of independent errors with mean 0 and variance σ^2. In a high-dimensional selection problem, β is supposed to be sparse, which means that most of the predictors do not have relationship with the response. The penalized likelihood is designed to identify which components of β are to be shrunk to zero and estimate the rest. The penalized likelihood estimator is

$$\hat{\beta} = \operatorname*{argmin}_{\beta} \left| y - \sum_{j=1}^{p} x_j \beta_j \right|^2 + \sum_{i=1}^{p} \rho(\lambda_j, |\beta_j|) \tag{7.1}$$

where $|\cdot|$ is the Euclidean norm, and $\rho(\lambda_j, .)$ is a given nonnegative penalty function that depends on the regularization parameter, λ_j. Tibshirani (1996) proposed the Lasso ℓ_1 penalty, with $\rho(\lambda_j, |\beta_j|) = \lambda|\beta_j|$. Fan and Li (2001) proposed a continuous differentiable penalty function whose derivative is defined by $(\partial/\partial u)\rho(\lambda_j, u) = \lambda\{1_{(u \leq \lambda)} + \{(a-1)\lambda\}^{-1}(a\lambda - u)_+ 1_{(u > \lambda)}\}$ for $a > 2$ and $u > 0$. The parameter a is shown to be fairly stable for the SCAD, and is usually set as 3.7 [Fan and Li (2001)]. Thus, for both Lasso and SCAD, one needs to determine λ in order to find an optimal model. It should be noted that the choice of λ is critically important because, for example, depending on the value of λ, it is possible to have none of the β's shrunk to zero, meaning that all of the predictors are selected, or all of the β shrunk to zero, meaning that none of the predictors are selected.

One approach to selecting regularization parameter is to use the information criteria such as the AIC [Akaike (1973)] and BIC [Schwarz (1978)]. Some data-driven approaches, such as the delete-d cross-validation [CV; Shao (1993)], the bootstrap model selection [Shao (1996)] and the generalised CV [Craven and Wahba (1979)] have also been proposed. More recent work can be found in Wang $et\ al.$ (2007) and Zhang $et\ al.$ (2010). These procedures are consistent if the regularization parameter satisfies some order conditions. However, there may be many functions satisfying these conditions and their finite sample behaviors may differ. Following the AF idea (see Section 3.1), Pang $et\ al.$ (2014) developed a data-driven approach to choosing the regularization parameter, as follows.

Let $\Lambda \subseteq [0, \infty)$ denotes the collection of candidate λ values. For each $\lambda \in \Lambda$, one chooses a model, $\hat{M}_\lambda$, by minimizing the penalized likelihood. Let $\mathcal{M}$ be the collection of all such $\hat{M}_\lambda, \lambda \in \Lambda$. Assume that the full model, M_f, is a correct model and included in $\mathcal{M}$. The first part of this assumption (i.e., a correct model is in $\mathcal{M}$) is necessary for consistent model selection [e.g., Jiang *et al.* (2008)]; the second part (i.e., the full model is in $\mathcal{M}$) is without loss of generality. Note that if both a correct model and the full model are in $\mathcal{M}$, the full model must be a correct model. However, usually, there are more than one correct models in $\mathcal{M}$. What one is interested is to find a correct model with minimum dimension, $M^* \in \mathcal{M}$, and this is known as the optimal model. For this purpose, one wishes to look for a λ that would maximize the probability of selecting M^*, with the Lasso or SCAD types of penalty functions, respectively. In other words, one wishes to select a λ that maximizes $P = \mathrm{P}(\hat{M}_\lambda = M^*)$. However, it is difficult, or even impossible, to compute the probability P theoretically for two reasons: First, the probability distribution P is unknown; and second, in practice, one does not know which model is optimal before the model selection.

A model-based bootstrap method is proposed to solve the first problem. Because the full model, M_f, is supposed to be a correct model, it is used to bootstrap the data. Let $r_i = y_i - x_i'\hat{\beta}_f$ be the ith residual, where $\hat{\beta}_f$ is the LS estimator under M_f. Let $\epsilon_1^*, \ldots, \epsilon_n^*$ be i.i.d. random variables generated from $N(0, \hat{\sigma}^2)$, where $\hat{\sigma}^2 = (n - p)^{-1} \sum_{i=1}^n r_i^2$. The bootstrap sample is $\{(y_i^*, x_i), i = 1, 2, \ldots, n\}$, where $y_i^* = x_i'\hat{\beta}_f + \epsilon_i^*$.

The maximized likelihood idea is used to solve the second problem. For each fixed λ, one generates B bootstrap samples. For each bootstrap sample, a model is selected by the Lasso (or SCAD). Let $n(M, \lambda)$ denote the number of times that model $M \in \mathcal{M}$ is selected with this λ out of the B bootstrap samples, and $p(M, \lambda) = n(M, \lambda)/B$ be the empirical probability that M is selected. Let $p(\lambda) = \max_{M \in \mathcal{M}} p(M, \lambda)$. In other words, $p(\lambda)$ is the largest empirical probability that a model is selected, given the λ. One then finds the λ that has the maximum $p(\lambda)$ as the optimal regularization parameter for the penalized likelihood.

A potential drawback of this method is that it may be computationally expensive even for a moderate B, say $B = 100$. For each $\lambda \in \Lambda$, a model is chosen by the Lasso (or SCAD) for each of the B bootstrap samples. If there are $|\Lambda| = d$ candidate λ's, the required Lasso or SCAD fit would have to be applied $B \times d$ times. One way to alleviate this problem is to use the efficient algorithm developed by Friedman *et al.* (2010) to solve the penalized likelihood based on coordinate descent. In addition, Pang *et al.*

(2014) proposed an accelerated algorithm to improve the computational efficiency. For each of the $\lambda \in \Lambda$, only $a\%$ of the B bootstrap samples are generated to calculate $p(\lambda)$. The λ's corresponding to the lowest $b\%$ of the $p(\lambda)$'s are removed from the candidate λ's. In other words, the new candidate λ's are the rest $(100 - b)\%$ of the original λ's. For each of the new candidate λ's, another $a\%$ of the B bootstrap samples are generated (the previous bootstrap samples can be reused to get more replications) and $b\%$ of the λ's are removed from the new candidates, and so on. It was shown that the accelerated algorithm is computationally much more efficient while performing nearly the same as the bootstrap algorithm without the acceleration in simulation studies.

Pang *et al.* (2014) studied performance of their procedure, which they called bootstrap selection (BS), in comparison with a number of existing methods of tuning parameter selection, including AIC, BIC, and CV, for Lasso, adaptive Lasso (ALasso), and SCAD. More specifically, to select λ via AIC (BIC), note that for any given λ, one can obtain a fitted model, say, $\hat{M}$, via Lasso (ALasso, SCAD), then evaluate the AIC (BIC) of $\hat{M}$. This way, the AIC (BIC) is a function of λ, and the optimal λ is the one that minimizes AIC (BIC). For CV, the regularization parameter is selected by the ten-fold cross-validation. A number of performance measures were considered. One is median of ratio of model error, defined as $\text{MRME} = \text{ME}(\hat{\lambda})/\text{ME}(0)$, where $\text{ME}(\lambda) = \{\hat{\beta} - \beta\}' \text{E}(xx')\{\hat{\beta} - \beta\}$, $\hat{\beta}$ being the shrinkage estimator under the tuning parameter λ and β being the true parameter vector, and $\hat{\lambda}$ is the selected optimal λ. Also, let C (IC) be the average number of correctly (incorrectly) identified nonzero coefficients; TP (true positive) be number of times, out of total of 100 simulation runs, that the identified nonzero coefficients are exactly those that are nonzero (no more, no less); UF be the number of times that at least one true nonzero coefficient is missed; and OF be the number of times that all of the nonzero true coefficients, plus at least one extraneous coefficient, are identified. We present the simulation results in subsequent examples.

Example 7.1. In this example, we consider a traditional fixed p situation. A similar setting has been used in Fan and Li (2001). The data are generated under the model $y_i = x_i'\beta + \epsilon_i, i = 1, \ldots, n$, where the true $\beta = (3, 1.5, 0, 0, 2, 0, 0, 0, 0, 0, 0, 0)'$ and ϵ_i's are generated independently from standard normal distribution. The predictors x_i's are i.i.d. normal vectors and with mean 0, variance 1, and correlation coefficient between the ith and jth elements being $\rho^{|i-j|}$ with $\rho = 0.5$. 100 simulated samples, each consist of $n = 100$ observations, were generated. In addition, a

Table 7.1 **Simulation Results for Example 7.1.**

Method	MRME	C	IC	UF	TP	OF
Lasso-BS	0.270	2.97	0.00	2	98	0
Lasso-CV	0.818	3.00	3.11	0	7	93
Lasso-AIC	0.803	3.00	2.91	0	12	88
Lasso-BIC	0.596	3.00	1.23	0	36	64
ALasso-BS	0.212	3.00	0.00	0	100	0
ALasso-CV	0.641	3.00	1.85	0	24	76
ALasso-AIC	0.648	3.00	1.49	0	40	60
ALasso-BIC	0.281	3.00	0.24	0	81	19
SCAD-BS	0.213	3.00	0.00	0	100	0
SCAD-CV	0.338	3.00	0.96	0	69	31
SCAD-AIC	0.598	3.00	1.59	0	41	59
SCAD-BIC	0.266	3.00	0.15	0	89	11
AIC	0.649	3.00	1.50	0	22	78
BIC	0.407	3.00	0.37	0	70	30

test data set of 10,000 observations were generated independently to compute the MRME. In addition to the shrinkage methods, the traditional AIC and BIC methods, based on all-subset selection, were also included in the comparison.

It can be seen that although all of the variable selection methods may reduce the model complexity and MRME, the BS method performs the best among all of the methods, with a very high TP (even reaching 100% when ALasso and SCAD are applied) and the MRME is quite small compared to the other methods. Also note that there is a small probability of UF by Lasso-BS, which happens when there are more than one λ^*'s. The CV performs better when used to select the regularization parameter for SCAD than for Lasso. Zhang *et al.* (2010) proved that when the true model is one of the candidate models, BIC is consistent under certain conditions while AIC is not. Thus, it is not surprising to see that BIC performs better than AIC in this situation. For the all-subset selection, BIC performs better than Lasso with BIC but worse than ALasso and SCAD with BIC. It is also found that ALasso improves Lasso significantly.

Example 7.2. The setup of this example is almost the same as Example 7.1, but a divergent number of parameters is considered. Let $\epsilon_i, i = 1, \ldots, n\}$ be generated independently from the normal distribution with mean 0 and standard deviation 6, and the true $\beta = (3 \cdot 1_q, 3 \cdot 1_q, 3 \cdot 1_q, 0_{p-3q})'$ with $p = p_n = [4n^{1/2}] - 5$, $q = [p_n/9]$, and 0_d being the d-dimensional vector of zeros. The sample sizes $n = 200$ and $n = 400$ were considered; thus, the corresponding p is 51 and 75, respectively, with the number of significant coefficients ($|\mathcal{A}|$) being 15 and 24, respectively (Zou and Zhang (2009)).

Table 7.2   **Simulation Results for Example 7.2.**

n	p_n	$|\mathcal{A}|$	Method	MRME	C	IC	UF	TP	OF
200	51	15	Lasso-BS	0.233	14.99	0.00	1	99	0
			Lasso-CV	0.628	15.00	9.46	0	0	100
			Lasso-AIC	0.641	15.00	9.89	0	0	100
			Lasso-BIC	0.424	15.00	2.79	3	13	87
			ALasso-BS	0.233	14.99	0.00	1	99	0
			ALasso-CV	0.556	15.00	6.30	0	5	95
			ALasso-AIC	0.625	15.00	6.68	0	6	94
			ALasso-BIC	0.354	15.00	1.29	3	36	61
			SCAD-BS	0.240	14.71	0.00	5	95	0
			SCAD-CV	0.600	15.00	7.62	0	4	96
			SCAD-AIC	0.620	15.00	36.93	0	6	94
			SCAD-BIC	0.396	14.95	2.98	3	21	76
400	75	24	Lasso-BS	0.296	23.54	0.00	2	98	0
			Lasso-CV	0.743	24.00	14.65	0	0	100
			Lasso-AIC	0.729	24.00	13.83	0	0	100
			Lasso-BIC	0.473	24.00	3.28	0	11	89
			ALasso-BS	0.296	24.00	0.00	0	100	0
			ALasso-CV	0.591	24.00	6.73	0	8	92
			ALasso-AIC	0.692	24.00	9.15	0	10	90
			ALasso-BIC	0.358	24.00	0.67	0	60	40
			SCAD-BS	0.296	23.53	0.00	4	96	0
			SCAD-CV	0.614	24.00	7.98	0	22	78
			SCAD-AIC	0.618	24.00	8.47	0	15	85
			SCAD-BIC	0.346	24.00	0.91	0	44	56

Table 7.2 summarizes the simulation results. In this case, it is difficult to apply the all-subset selection of AIC and BIC (for example, even when $p = 51$, there are a total of $2^{51} - 1$ subsets). The BS method still performs the best with the probability of TP tending to one. The CV and AIC with all the penalities almost lose their power with the probability of TP tending to zero. BIC with Lasso only performs slightly better than those with AIC. BIC with ALasso and SCAD performs better but is still much worse than BS in these sparse settings.

Example 7.3. We now consider a so-called ultra-high dimensional situation in which $p \gg n$, namely, $n = 200$ and $p = p_n = 1000$. This is an example from citetfan08 and Zou and Zhang (2009). The data were generated under the model $y_i = x'_i \beta + \epsilon_i$, where the true $\beta = [\beta'_{(1)}, 0_{p_n - |\mathcal{A}|}]'$ with $|\mathcal{A}| = 8$; $\beta_{(1)}$ is a vector of size 8 with each component being of the form $(-1)^u(a_n + |z|)$, where $a_n = 4\log(n)/n^{1/2}$, u is generated from the Bernoulli(0.4) distribution, and z from the $N(0, 1)$ distribution; and $\epsilon_i, i = 1, \ldots, n$ are generated independently from the $N(0, 1.5^2)$. The components of the predictors, $x_i, i = 1, \ldots, n$, are independent standard normal.

Table 7.3 **Simulation Results for SIS Plus Various Penalities.**

n	p_n	$|\mathcal{A}|$	Method	C	IC	UF	TP	OF
200	1000	8	SIS+Lasso-BS	7.99	0.00	1	99	0
			SIS+Lasso-BIC	8.00	14.87	0	1	99
			SIS+ALasso-BS	7.99	0.00	1	99	0
			SIS+ALasso-BIC	8.00	2.62	0	37	63
			SIS+SCAD-BS	8.00	0.00	0	100	0
			SIS+SCAD-BIC	8.00	1.48	0	62	38

First, using the sure independence screening [SIS; Fan and Lv (2008)] method, the dimensionality of the model was reduced from 1000 to $n_{\text{sis}} = [5.5n^{2/3}] = 188$, before any penalized likelihood method was applied. Note that 188 is only slightly smaller than the sample size n. From the previous examples, it seems reasonable to only compare the BIC based methods. The results are summarized in Table 7.3. The BS method is still capable of selecting exactly the correct model with probability of TP equal or very close to one. The Lasso-BIC totally loses the power and the ALasso-BIC and SCAD-BIC are much inferior to the BS.

7.2 Shrinkage variable selection for GLM

Generalized linear models (GLM; McCullagh and Nelder (1989)) extend the notion of linear models to situations of discrete or categorical responses, such as binomial or Poisson observations. To define a GLM, let us first consider an alternative expression of the Gaussian linear model, in which the observations, $y_1, \ldots, y_n$ are independent such that $y_i \sim N(x_i'\beta, \tau^2)$, where x_i is a known vector of covariates, β is an unknown vector of regression coefficients, and τ^2 is an unknown variance. The two key elements in the above that define a Gaussian linear model are (i) independence of the data; and (ii) normal distribution. Element (i) is kept for the GLM. As for (ii), the normal distribution is replaced by a distribution that is a member of the exponential family with pdf

$$f_i(y_i) = \exp\left\{\frac{y_i\xi_i - b(\xi_i)}{a_i(\phi)} + c_i(y_i, \phi)\right\}, \tag{7.2}$$

where $b(\cdot)$, $a_i(\cdot)$, $c_i(\cdot, \cdot)$ are known functions, and ϕ is a dispersion parameter, which may or may not be known. We shall add one more element. According to properties of the exponential family [McCullagh and Nelder (1989)], the quantity ξ_i is naturally associated with the mean $\mu_i = \text{E}(y_i)$.

We assume that (iii) μ_i, in turn, is associated with a linear predictor

$$\eta_i = x_i'\beta, \tag{7.3}$$

where x_i and β play similar roles as in the Gaussian linear model, through a known link function $g(\cdot)$ such that

$$g(\mu_i) = \eta_i. \tag{7.4}$$

It is easy to verify that the Gaussian linear model is a special case of GLM with the link function being the identity function, that is, $g(\mu) = \mu$.

There have been extensions of the shrinkage methods for linear models to GLM. See, for example, Park and Hastie (2007), who addresses the computational issue by introducing a regularization path algorithm for Lasso in GLM (see below), in a similar fashion to the LARS algorithm [Efron *et al.* (2004)]. A more general regularization path, which can be used for Lasso, a ridge penalty, and the elastic net [Zou and Hastie (2005)] focusing on sparse data has also been implemented [Friedman *et al.* (2010)]. Shrinkage estimators in the form of a group Lasso for logistic regression has been implemented by Meier *et al.* (2008). On the other hand, there has not been much focus on selection of the regularization parameter in GLM in the current literature. An extension of the shrinkage estimation to GLM is defined by minimizing a penalized negative log-likelihood function, which has the form

$$-\log\{L(\beta|y)\} + \lambda\rho(\beta), \tag{7.5}$$

where $L(\beta|y)$ is the likelihood function for estimating β given the data y, and $\rho(\cdot)$ is a penalty function. Here we assume, for simplicity, that the dispersion parameter, ϕ, is known. The latter case includes the Binomial and Poisson distributions, as special (and perhaps the most popular) cases. For example, in the special case of logistic regression, one has

$$\log\{L(\beta|y)\} = \sum_{i=1}^{n}\{y_i\eta_i - \log(1 + e^{\eta_i})\} \tag{7.6}$$

with η_i given by (7.4). The discussion below is based on Melcon (2014), which focuses on the above special case (although the same idea can be generalized), and two kinds of penalties: (I) $\rho(\beta) = \sum_{j=1}^{d}|\beta_j|$ (the Lasso penalty), where d is the dimension of β, and (II) $\rho(\beta) = \sum_{j=1}^{d}|\beta_j|/|\hat{\beta}_j|$ (the adaptive Lasso penalty). The goal is, again, to select the regularization parameter, λ, that correctly identifies the optimal model, here defined as a true model with the minimum dimension. The idea is similar to that for the linear model case. Let $\mathcal{M}$ be the set of all candidate models, M^* be the unknown optimal model, which is assumed to belong to $\mathcal{M}$, and M be a candidate model. For a given λ, we select a model $\hat{M}^\lambda$ via the shrinkage estimation. We wish to find the value of λ that maximizes the empirical probability that $\hat{M}^\lambda$ is the optimal model, M^*. That is, we want to maximize $p = P(\hat{M}^\lambda = M^*)$. Since we do not, in general, know the distribution of P, and we do not know the optimal model, we cannot calculate the probability p exactly. Instead, we estimate p empirically using bootstrap samples. Let $y = (y_i)_{1 \le i \le n}$ be the vector of responses, and $X = (x'_i)_{1 \le i \le n}$ the matrix of covariates. Let M_{f} denote the full model that includes all of the candidate covariate variables. Let the MLE of β under M_{f} be $\hat{\beta}_{\mathrm{f}}$. We draw a parametric bootstrap sample, $y^{(b)}$, under M_{f}, treating $\hat{\beta}_{\mathrm{f}}$ as the true parameters. For each value of λ among a grid of candidate values, and each bootstrap sample $y^{(b)} = [y_i^{(b)}]_{1 \le i \le n}$, $1 \le b \le B$, we minimize (7.5) with y_i replaced by $y_i^{(b)}$, $1 \le i \le n$. We then count the frequency over the bootstrap samples that each model, M, is selected, denoted by $f^\lambda(M)$. Note that the selected model corresponds to the subset of estimated β coefficients that are not shrunk to zero. The relative frequency is then $p^\lambda(M) = f^\lambda(M)/B$. Let $p^\lambda = \max_{M \in \mathcal{M}} p^\lambda(M)$. This results in a p^λ for every value of λ among the grid. To pick the overall optimal λ, it is desirable to plot p^λ vs λ, and choose the λ corresponding to the first high "peak" in the plot. The method is called called empirical optimal selection (EOS) by Melcon (2014).

Notes: Note that at $\lambda = 0$, $p^\lambda = 1$, because the unpenalized (or full) model will be chosen for every bootstrap sample. This λ is removed from the possible candidates. Similarly, when λ is sufficiently large, $p^\lambda = 1$, because all coefficients will be shrunk to zero and the empty model will be chosen for every bootstrap sample (Exercise 7.2). This is clearly marked by a horizontal line at the right edge of the plot in Figure 7.1. These λ's are also removed from the possible candidates. Also, it is possible to have two high peaks in the plot. See, for example, Figure 7.2. In this case, it is recommended to chose the λ corresponding to the first peak as the best choice, as this leads to a more conservative choice in model selection. Finally, when using adaptive Lasso, the weights $\hat{\beta}$ are fixed as $\hat{\beta}_{\mathrm{f}}$ for all bootstraps. The justification is that one is selecting a model based on the original data X, y (while bootstrapping is simply considered a way of computation), and the weights $\hat{\beta}_{\mathrm{f}}$ reflect information from the original data.

Once again, we demonstrate the performance of the EOS method, with comparison with other procedures, through a number of simulated exam-

Table 7.4 **Simulation Results for GLM/Lasso.**

	TP	OF	UF	C	IC	$\bar{\lambda}$ (s.d.)
			$n = 100$			
EOS	65	20	15	2.71	0.24	1.271 (1.022)
AIC	9	91	0	3	1.65	0.121(0.1)
BIC	35	63	2	2.97	0.93	0.267 (0.21)
BIC_γ	61	24	15	2.83	0.28	0.499 (0.308)
CV	5	95	0	3	1.77	0.202 (0.111)
			$n = 300$			
EOS	100	0	0	3	0	1.393 (0.23)
AIC	20	80	0	3	1.52	0.125 (0.094)
BIC	51	49	0	3	0.64	0.3 (0.156)
BIC_γ	78	22	0	3	0.25	0.425 (0.191)
CV	17	83	0	3	1.58	0.198 (0.114)

ples. The results are copied from Melcon (2014).

Example 7.4. For the first example, the true value of β is set $\beta = (2, 0, 2, 2, 0, 0)'$ (of dimension 6 with 3 significant predictors), and n be either 100 or 300. In addition to EOS, several competing methods of choosing the regularization parameter are considered, which include AIC, BIC and CV, as in the previous section, as well as the BIC_γ method proposed by Chen and Chen (2008). The latter replaces the BIC penalty, $\log(n)d_\lambda$, where d_λ is the number of nonzero β estimates for the given λ, by $d_\lambda \log(n) + 2\gamma \log(C_{d_\lambda}^p)$ with C_d^p being the binomial coefficient of "p choose d, and $\gamma = 0.5$ if $p < n$; $\gamma = 1$ if $p > n$ [Chen and Chen (2008)]. The results are presented in Tables 7.4 and 7.5. The measures of performance are defined the same way as those in Tables 7.1–7.3. In addition, the mean ($\bar{\lambda}$) and s.d. of the selected λ, based on the simulation results, are provided. Some sample plots of the empirical probabilities based on the simulation results are presented in Figures 7.1 and 7.2 for illustrations.

Example 7.5. To examine performance of EOS when the dimension of β increases, another simulation study was carried out with the true $\beta = (2, 0, 0, 0, 2, 0, 2, 0, 2, 0)'$ (dimension 10 with 4 significant predictors) and otherwise repeating the settings of Example 7.4. The results are presented in Tables 7.6 and 7.7.

Table 7.5 **Simulation Results for GLM/ALasso.**

	TP	OF	UF	C	IC	$\bar{\lambda}$ (s.d.)
			$n = 100$			
EOS	70	21	9	2.91	0.27	1.212 (1.025)
AIC	40	60	0	3	0.87	0.08 (0.066)
BIC	65	31	4	2.96	0.38	0.176 (0.133)
BIC_γ	78	8	14	2.85	0.1	0.313 (0.267)
CV	46	50	4	2.96	0.83	0.155 (0.142)
			$n = 300$			
	TP	OF	UF	C	IC	$\bar{\lambda}$ (s.d.)
EOS	100	0	0	3	0	1.952 (0.763)
AIC	52	48	0	3	0.63	0.056 (0.044)
BIC	89	11	0	3	0.11	0.129 (0.096)
BIC_γ	96	4	0	3	0.04	0.151 (0.11)
CV	67	33	0	3	0.57	0.099 (0.08)

Table 7.6 **Simulation Results for GLM/Lasso.**

	TP	OF	UF	C	IC	$\bar{\lambda}$ (s.d.)
			$n = 100$			
EOS38	38	24	3.63	1	0.974 (0.777)	
AIC	4	96	0	4	3.61	0.121 (0.093)
BIC	22	67	11	3.82	1.65	0.375 (0.285)
BIC_γ	37	29	34	3.35	0.49	0.715 (0.441)
CV	38	56	6	3.93	1.36	0.186 (0.128)
			$n = 300$			
	TP	OF	UF	C	IC	$\bar{\lambda}$ (s.d.)
EOS	84	14	2	3.96	0.32	1.679 (6.853)
AIC	6	94	0	4	3.42	0.142 (0.094)
BIC	36	64	0	4	1.2	0.372 (0.156)
BIC_γ	61	39	0	4	0.53	0.508 (0.188)
CV	51	49	0	4	0.97	0.12 (0.059)

7.3 Connection to the fence

As noted, the methods of Pang *et al.* (2014) and Melcon (2014) were motivated by the fence methods. In fact, the similarity between BS (Section 7.1) and EOS (Section 7.2) and AF (Section 3.1) is apparent. We now point out that, not only the BS and EOS are similar to the AF, they are, indeed, special cases of the AF.

To see this, let $\lambda(M)$ be the smallest regularization parameter, λ, for which model M is selected. It is easy to see that $\lambda(M)$ is a measure of lack-of-fit in the sense that "larger" model has smaller $\lambda(M)$. More specifically, if M_1 is a submodel of M_2, then, one must have $\lambda(M_1) \geq \lambda(M_2)$ (Exercise

Probability Plot for Fixed Effects

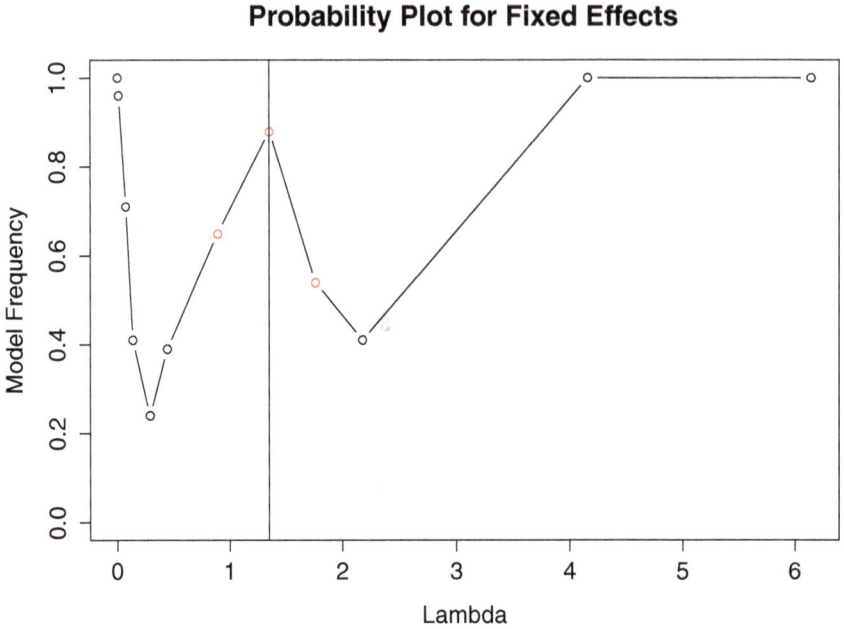

Fig. 7.1   *Sample Plot for GLM/Lasso with* $n = 300$.

7.3). In particular, we have $\lambda(M_f) = 0$, where M_f is the full model. Thus, if we define $Q(M) = \lambda(M)$ in (3.2), the fence inequality becomes

$$\lambda(M) \leq c. \tag{7.7}$$

It follows that the selection of the optimal λ is equivalent to selecting the optimal cut-off, c in (7.7). Thus, the BS procedure of Pang *et al.* (2014) and EOS procedure of Melcon (2014) may be viewed as special cases of the AF with $Q(M) = \lambda(M)$. Therefore, the BS and EOS can be justified by the AF, whose asymptotic properties such as consistency in model selection, in particular, are considered in Chapter 9.

7.4   Shrinkage mixed model selection

There has been some recent work on joint selection of the fixed and random effects in mixed effects models. Bondell *et al.* (2010) considered such a selection problem in a certain type of linear mixed models, which can be

Probability Plot for Fixed Effects

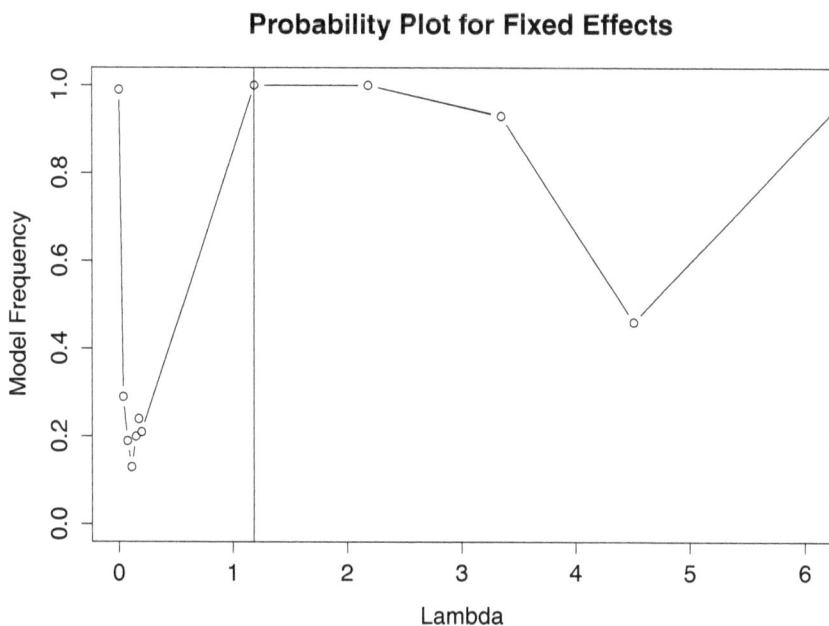

Fig. 7.2 *Sample Plot for GLM/Adaptive Lasso with $n = 300$.*

expressed as

$$y_i = X_i\beta + Z_i\alpha_i + \epsilon_i, \quad i = 1,\ldots,m, \tag{7.8}$$

where y_i is an $n_i \times 1$ vector of responses for subject i, X_i is an $n_i \times p$ matrix of explanatory variables, β is a $p \times 1$ vector of regression coefficients (the fixed effects), Z_i is an $n_i \times q$ known design matrix, α_i is a $q \times 1$ vector of subject-specific random effects, ϵ_i is an $n_i \times 1$ vector of errors, and m is the number of subjects. It is assumed that the $\alpha_i, \epsilon_i, i = 1,\ldots,m$ are independent with $\alpha_i \sim N(0, \sigma^2\Psi)$ and $\epsilon_i \sim N(0, \sigma^2 I_{n_i})$, where Ψ is an unknown covariance matrix. The problem of interest, using the terms of shrinkage model selection, is to identify the nonzero components of β and $\alpha_i, 1 \le i \le m$. For example, the components of α_i may include a random intercept and some random slopes.

To take advantage of the idea of shrinkage variable selection, Bondell *et al.* (2010) adopted a modified Cholesky decomposition. Note that the covariance matrix, Ψ, can be expressed as $\Psi = D\Omega\Omega'D$, where $D = \text{diag}(d_1,\ldots,d_q)$ and $\Omega = (\omega_{kj})_{1 \le k, j \le q}$ is a lower triangular matrix

Table 7.7 **Simulation Results for GLM/ALasso.**

	TP	OF	UF	C	IC	$\bar{\lambda}$ (s.d.)
			$n = 100$			
EOS	59	29	12	3.86	0.59	1.192 (0.941)
AIC	18	81	1	3.99	1.9	0.1 (0.062)
BIC	45	48	7	3.93	0.73	0.243 (0.147)
BIC_γ	51	21	28	3.59	0.28	0.469 (0.364)
CV	36	55	9	3.9	1.28	0.192 (0.124)
			$n = 300$			
	TP	OF	UF	C	IC	$\bar{\lambda}$ (s.d.)
EOS	96	3	1	3.98	0.03	1.199 (0.474)
AIC	36	64	0	4	1.29	0.077 (0.048)
BIC	74	26	0	4	0.3	0.161 (0.081)
BIC_γ	85	15	0	4	0.15	0.195 (0.099)
CV	56	44	0	4	0.77	0.128 (0.074)

with 1's on the diagonal. Thus, one can express (7.8) as

$$y_i = X_i\beta + Z_i D\Omega\xi_i + \epsilon_i, \quad i = 1, \ldots, m, \qquad (7.9)$$

where the ξ_i's are independent $N(0, \sigma^2)$ random variables. The idea is to apply shrinkage estimation to both $\beta_j, 1 \leq j \leq p$ and $d_k, 1 \leq k \leq q$. Note that setting $d_k = 0$ is equivalent to setting all of the elements in the kth column and kth row of Ψ to zero, and thus creating a new submatrix by deleting the corresponding row and column, or the exclusion of the kth component of α_i. However, direct implementation of this idea is difficult, because the ξ_i's are unobserved, even though their distribution is much simpler. To overcome this difficulty, Bondell *et al.* (2010) used the E-M algorithm [Dempster *et al.* (1977)]. By treating the ξ_i's as observed, the complete-data log-likelihood can be expressed as (Exercise 7.4)

$$l_c = c_0 - \frac{N + mq}{2}\log\sigma^2 - \frac{1}{2\sigma^2}(|y - X\beta - Z\tilde{D}\tilde{\Omega}\xi|^2 + |\xi|^2), \quad (7.10)$$

where c_0 is a constant, $N = \sum_{i=1}^{m} n_i$, $X = (X_i)_{1 \leq i \leq m}$, $Z = \text{diag}(Z_1, \ldots, Z_m)$, $\tilde{D} = I_m \otimes D$, $\tilde{\Omega} = I_m \otimes \Omega$ ($\otimes$ means Kronecker product), $\xi = (\xi_i)_{1 \leq i \leq m}$, and $|\cdot|$ denotes the Euclidean norm. (7.10) leads to the shrinkage estimation by minimizing

$$P_c(\phi|y,\xi) = |y - X\beta - Z\tilde{D}\tilde{\Omega}\xi|^2 + \lambda\left(\sum_{j=1}^{p}\frac{|\beta_j|}{|\tilde{\beta}_j|} + \sum_{k=1}^{q}\frac{|d_j|}{|\tilde{d}_j|}\right), \quad (7.11)$$

where ϕ represents all of the parameters, including the β's, the d's, and the ω's, $\tilde{\beta} = (\tilde{\beta}_j)_{1 \leq j \leq p}$ is given by the right side of (6.15) with the variance components involved in $V = \text{Var}(y)$ replaced by their REML estimators [e.g.,

Jiang (2007), sec. 1.3.2], $\tilde{d}_k, 1 \le k \le q$ are obtained by decomposition of the estimated Ψ via the REML, and λ is the regularization parameter. Bondell *et al.* (2010) proposed to use the BIC in choosing the regularization parameter. Here the form L^1 penalty in (7.11) is in terms of the adaptive Lasso [Zou (2006)]. To incorporate with the E-M algorithm, one replaces (7.11) by its conditional expectation given y, and the current estimate of ϕ. Note that only the first term on the right side of (7.11) involves ξ, with respect to which the conditional expectation is taken. The conditional expectation is then minimized with respect to ϕ to obtain the updated (shrinkage) estimate of ϕ. A similar approach was taken by Ibrahim *et al.* (2011) for joint selection of the fixed and random effects in GLMMs [e.g., Jiang (2007)], although the performance of the proposed method was studied only for the special case of linear mixed models.

As in Sections 7.1 and 7.2, one may use the AF idea to derive a data-driven approach to selection of the regularization parameters in shrinkage mixed model selection, such as the λ in (7.11). Below we consider selection problems from a different perspective.

Suppose that the purpose of the joint selection is for predicting some mixed effects, as discussed in Section 6.4. We may combine the predictive measure of lack-of-fit developed in Section 6.4 with the shrinkage idea of Bondell *et al.* (2010) and Ibrahim *et al.* (2011). Namely, we replace the first term on the right side of (7.11) by the predictive measure,

$$Q(\phi|y) = (y - X\beta)'\Gamma\Gamma'(y - X\beta) - 2\text{tr}(\Gamma'\Sigma) \tag{7.12}$$

[see the paragraph above Example 6.2 (continued) that involved (6.18)]. The regularization parameter, λ, may be chosen using a similar procedure as in Sections 7.1 and 7.2. We refer to this method as *predictive shrinkage selection*, or PSS. The idea has been explored by Hu *et al.* (2015). The authors found that PSS performs better than the shrinkage selection method based on (7.11) not only in terms of the predictive performance, but also in terms of *parsimony*. The latter refers to the classical criterion of selecting a correct model with the minimum number of parameters. The authors also extended the PSS to Poisson mixed models, a special case of GLMM, and obtained similar results. There is another advantage of the PSS in terms of computation. Denote the first term on the right side of (7.11) by $Q_c(\phi|y,\xi)$. Note that, unlike $Q_c(\phi|y,\xi)$, the unobserve ξ is not involved in $Q(\psi|y)$ defined by (7.12), which is what the PSS is based on. This means that, unlike Bondell *et al.* (2010) and Ibrahim *et al.* (2011), PSS does not need to run the E-M algorithm, and thus is computationally (much) more

Table 7.8 **Predictors for Diabetes Data.**

age	sex	bmi	map	tc	ldl	hdl	tch	ltg	glu
1	2	3	4	5	6	7	8	9	10

Table 7.9 **Variable Selection for Diabetes Data.**

Method	Model	Method	Model	Method	Model
AIC	(2,3,4,5,6,9)	BIC	(2,3,4,7,9)		
Lasso-BS	(3,9)	SCAD-BS	(3,9)	ALasso-BS	(3,9)
Lasso-AIC	(2,3,4,5,7,9,10)	SCAD-AIC	(2,3,4,5,6,9)	ALasso-AIC	(2,3,4,5,6,8,9)
Lasso-BIC	(2,3,4,5,7,9,10)	SCAD-BIC	(2,3,4,5,6,9)	ALasso-BIC	(2,3,4,5,6,8,9)
Lasso-CV	(2,3,4,5,7,9,10)	SCAD-CV	(2,3,4,5,6,9)	ALasso-CV	(2,3,4,5,6,8,9,10)

efficient. We illustrate this with a real-data example in the next section.

7.5   Real data examples

7.5.1   *Diabetes data*

Efron *et al.* (2004) used a data set from a diabetes study involving 442 patients measured on 10 baseline variables, listed in Table 7.8, as well as the response variable which is a measure of disease progression one year after the baseline. The baseline variables are considered as potential predictors. The authors used the data to illustrate the LARS.

We use the same data to illustrate the BS method discussed in Section 7.1. The result is presented in Table 7.9 and compared with those using AIC, BIC, and CV for choosing the regularization parameter. For each penalty function, AIC, BIC and CV gives almost the same model but the selected models are different for different penalties. In contrast, the BS method chooses the same model for all of the penalties. The selected model is simpler than those by the other methods. Scatter plots of the data (omitted) seem to suggest that some of the predictors are correlated, which may explain why some of the predictors are dropped by BS than by the other methods. Also note that the selection is restricted to linear models; so, even if some variables are not selected, it only means that they are not considered to have linear association with the response. Nonlinear associations, however, may still exist.

7.5.2   *Heart disease data for South African men*

Hastie *et al.* (2001) considered a data set on heart disease in South African

Table 7.10 **South African Predictors.**

Coefficient (Variable Name)	Short Description
β_1 (sbp)	Systolic Blood Pressure
β_2 (tobacco)	Cumulative Tobacco (kg)
β_3 (ldl)	Low Density Lipoprotein Cholesterol
β_4 (adiposity)	Amount of Fat Found in Adipose Tissue
β_5 (famhist)	Family History of Heart Disease
β_6 (typea)	Type-A Behavior rating
β_7 (obesity)	Obesity Score
β_8 (alcohol)	Current Alcohol Consumption
β_9 (age)	Age at Onset

Table 7.11 **Variable Selection via Lasso for South African Data.**

	β_0	β_1	β_2	β_3	β_4	β_5	β_6	β_7	β_8	β_9
EOS	X		X	X		X	X			X
AIC	X	X	X	X		X	X	X		X
BIC	X	X	X	X		X	X	X		X
BIC_γ	X	X	X	X		X	X			X
CV	X	X	X	X		X	X	X		X

Table 7.12 **Variable Selection via Adaptive Lasso for South African Data.**

	β_0	β_1	β_2	β_3	β_4	β_5	β_6	β_7	β_8	β_9
EOS	X		X	X		X	X			X
AIC	X		X	X		X	X	X		X
BIC	X		X	X		X	X			X
BIC_γ	X		X	X		X	X			X
CV	X	X	X	X		X	X	X		X

men. The responses are binary indicates of whether or not the person had heart disease. The data involves a total of 462 observations. A number of potential predictors are considered, which are listed in Table 7.10.

We use this example to illustrate the EOS method discussed in Section 7.2. The EOS was applied to the data with the empirical probability plot shown in Figure 7.3. The plot shows a clear "first peak", which corresponds to the choice for the optimal λ. Results of the EOS model selection are presented in Tables 7.11 and 7.12, and compared with results of shrinkage model selection with the λ chosen by the (ten-fold) CV, AIC, BIC, and BIC_γ. An X indicates that the predictor corresponding to the β coefficient was included (and a blank indicates that the corresponding predictor was excluded). To evaluate the performances of different methods empirically, 10-fold cross validation (c.v.) was used to assess the predictive power of

SA data lasso

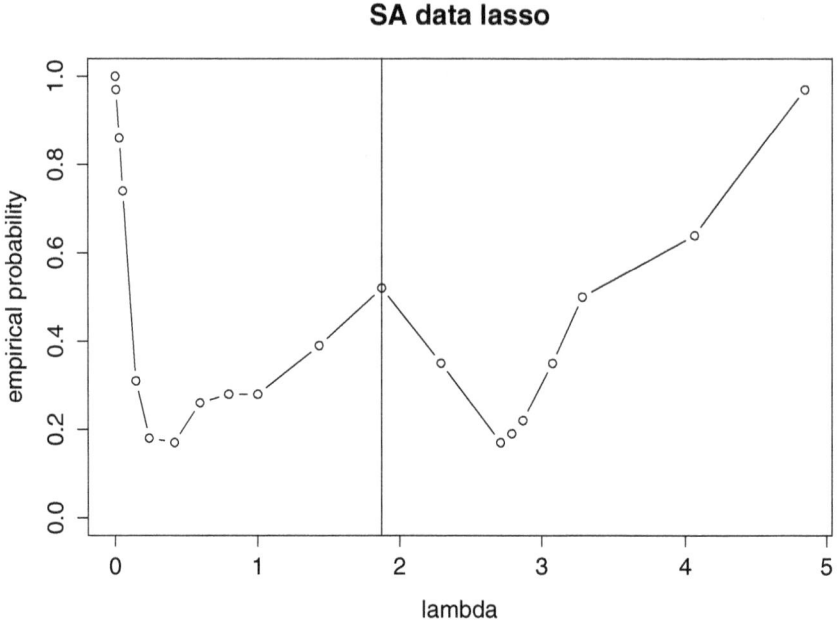

Fig. 7.3 *Empirical Probability Plot for South African Data.*

the selected model. The c.v. was implemented 100 times. The boxplots
of the resulting MSPE values are presented in Figure 7.4. It appears that,
when using the Lasso penalty, EOS selects the model with the smallest
MSPE. In the case of adaptive Lasso, EOS, BIC, and BIC_γ all choose
the same model, and their MSPE's are comparable (unsurprisingly; plot
omitted). This follows a general trend seen in the simulations, that is,
different methods find more agreement when using adaptive Lasso, while
EOS tends to choose a different model when using Lasso.

7.5.3   *Analysis of high-speed network data*

Finally, we consider an application of the PSS method, introduced in Section 7.4. Efficient data access is essential for sharing massive amounts of
data among geographically distributed research collaborators. The analysis
of network traffic is getting more and more important today for utilizing
limited resources offered by the network infrastructures and planning wisely
large data transfers. The latter can be improved by learning the current
conditions and accurately predicting future network performance. Short-

MSPE for lasso

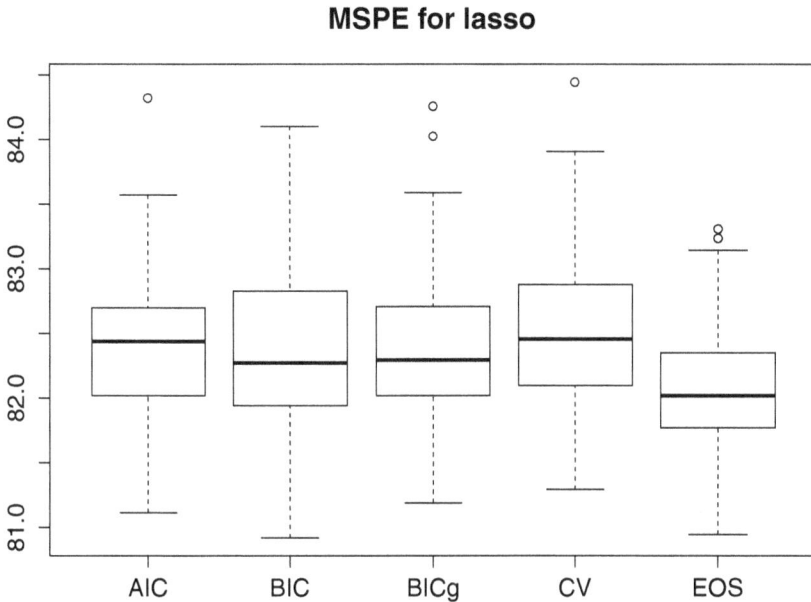

Fig. 7.4 *Boxplots of Empirical MSPE for South African Data.*

term prediction of network traffic guides the immediate scientific data place-
ments for network users; long-term forecast of the network traffic enables
capacity-planning of the network infrastructure needs for network designers. Such prodictions become non-trivial when the amount of network data
grows in unprecedented speed and volumes. One such avialable data source
is NetFlow [Cisco Systems (1966)].

The NetFlow measurements provide high volume, abundant specific information for each data flow; some sample records are shown in Table 7.13
(with IP addresses masked for privacy). For each record contains the following list of variables:

Start, End The start and end time of the recorded data transfer.
Sif, Dif The source and destination interface assigned automatically for
the transfer.
SrcIPaddress, DstIPaddress The source and destination IP addresses
of the transfer.
SrcP, DstP The source and destination Port chosen based on the transfer
type such as email, FTP, SSH, etc.

Table 7.13 **Sample NetFlow Records.**

| Start | End | Sif | SrcIPaddress(masked) | SrcP | Dif |
DstIPaddress(masked)	DstP	P	Fl	Pkts	Octets
0930.23:59:37.920	0930.23:59:37.925	179	xxx.xxx.xxx.xxx	62362	175
xxx.xxx.xxx.xxx	22364	6	0	1	52
0930.23:59:38.345	0930.23:59:39.051	179	xxx.xxx.xxx.xxx	62362	175
xxx.xxx.xxx.xxx	28335	6	0	4	208
1001.00:00:00.372	1001.00:00:00.372	179	xxx.xxx.xxx.xxx	62362	175
xxx.xxx.xxx.xxx	20492	6	0	2	104
0930.23:59:59.443	0930.23:59:59.443	179	xxx.xxx.xxx.xxx	62362	175
xxx.xxx.xxx.xxx	26649	6	0	1	52
1001.00:00:00.372	1001.00:00:00.372	179	xxx.xxx.xxx.xxx	62362	175
xxx.xxx.xxx.xxx	26915	6	0	1	52
1001.00:00:00.372	1001.00:00:00.372	179	xxx.xxx.xxx.xxx	62362	175
xxx.xxx.xxx.xxx	20886	6	0	2	104

P The protocol chosen based on the general transfer type such as TCP, UDP, etc.

Fl The flags measured the transfer error caused by the congestion in the network.

Pkts The number of packets of the recorded data transfer.

Octets The Octets measures the size of the transfer in bytes.

Features of NetFlow data have led to consideration of GLMMs (e.g., Jiang 2007) for predicting the network performance. First, NetFlow record is composed of multiple time series with unevenly collected time stamps. Because of this feature, traditional time series methods such as ARIMA model, wavelet analysis, and exponential smoothing [e.g., Fan and Yao (2003), sec. 1.3.5] encounter difficulties, because these methods are mainly designed for evenly collected time stamps and dealing with a single time series. Thus, there is a need for modeling a large number of time series without constraints on even collection of time stamps. On the other hand, GLMM is able to fully utilize all of the variables involved in the data set without requiring evenly-spaced time variable.

Second, there are empirical evidences of associations, as well as heteroscedasticity, found in the NetFlow records. For example, with increasing number of packets in a data transfer, it takes longer, in general, to finish the transfer. This suggests that the number of packets may be considered as a (fixed) predictor for the duration of data transfer. Furthermore, there appear to be fluctuation among network paths in terms of slope and range in the plots of duration against the number of packets. See Fig. 1 of Hu *et al.* (2014). This suggests that the network path for data transfer may be associated with a random effect to explain the duration under varying conditions. Thus, again, a mixed effects model seems to be plausible. Moreover, GLMM is more flexible than linear mixed model in terms of the mean-variance association.

Third, NetFlow measurements are big data with millions of observation for a single router within a day and 14 variables in each record with 30s or 40s interaction terms as candidates. The large volume and complexity of the data require efficient modeling. However, this is difficult to do with fixed effects modeling. For example, traditional hierarchical modeling requires dividing the data into groups, but the grouping is not clear, and requires investigation to identify the variable that classify the observed data. Explorative data analysis shows that the grouping factor could be the path of the data transfer, the delivering time of the day, the transfer protocol used, or the combination of some or all of these. With so many uncertainties, one approach to simplifying the modeling is via the use of random effects [Jiang (2007)].

The NetFlow data used for the current analysis was provided by ESnet for the duration from May 1, 2013 to June 30, 2013. Considering the network users' interests, the established model should be able to predict the duration of a data transfer so that the users can expect how long it would take for the data transfer, given the size of their data, the start time of the transfer, selected path and protocols. Considering the network designers' interests, the established model should be able to predict the long-time usage of the network so that the designer will know which link in the network is usually congested and requires more bandwidth, or rerouting of the path. In the following, we illustrate a model built for these interests, and compare its prediction accuracy with two traditional GLMM procedures: Backward-Forward selection (B-F) and Estimation-based Lasso (E-Lasso). The latter refers to Lasso for the penalized likelihood method, corresponding to (7.11) [Bondell *et al.* (2010)].

The full model predicts the transfer duration, assuming influences from the fixed effects including transfer start time, transfer size (Octets and Packets) and the random effects including network transfer conditions such as Flag and Protocol, source and destination Port numbers, and transfer path such as source and destination IP addresses and Interfaces. The PSS procedure, with the Lasso penalty (see Section 7.4), has selected the model

$$y = \beta_{\text{start}} s(x_{\text{start}}) + \beta_{\text{pkt}} x_{\text{pkt}} + Z_{\text{ip-path}} v_{\text{ip-path}} + e, \qquad (7.13)$$

where $s(\cdot)$ is a fitted smoothing spline implemented to taken into account that the mean response is usually nonlinearly associated with the time

Table 7.14 **Estimates of Non-zero Fixed Effects.**

Fixed Effects	Estimates	Standard Error	P-value
Intercept	-13.809	0.914	<2e-16
Start Time	0.574	0.0169	<2e-16
Packets	1.115	0.035	<2e-16

Table 7.15 **Comparison of PSS, B-F, E-Lasso in Term of MSPE and Computing Time.**

	E-Lasso	B-F	PSS
MSPE	2306	42230	127
Time (in seconds)	6.26×10^7	5.43×10^{10}	142

variable, x_{start}, with the parameters of the smoothing spline chosen automatically by cross-validation. The parameter estimates and their corresponding P-values for model (7.13) are given in Table 7.14. The fitted smoothing spline is plotted in Figure 7.5, which shows how the transfer duration varies with start time. Furthermore, in (7.13), $Z_{\text{ip-path}}$ is the design matrix whose columns correspond to the ip-paths, and $v_{\text{ip-path}}$ is a vector-valued random effect whose components correspond to the ip-paths, and e is an additional error corresponding to the background noise. The PSS has identified six paths with non-zero random-effect standard deviations, indexed as 14, 16, 38, 41, 61, and 83. Among those paths, all except path 61 have the estimated standard deviation of at least 10, while the estimated standard deviation for path 61 is almost zero. A plot is shown in Figure 7.6. The estimated standard deviation for the background noise is 11.239.

Regarding the comparison of PSS with B-F and E-Lasso, the overall MSPE as well as computation speed for the comparing methods are presented in Table 7.15. Note that, in this case, one actually knows the truth for the prediction, and therefore is able to compute the (exact) MSPE, and record the total computational time, of course. The results show that, in terms of prediction accuracy (MSPE), PSS is about 18 times better than E-Lasso, and 330 times better than B-F; in terms of computational time, PSS is about 4×10^5 times less than E-Lasso and 3.8×10^8 times less than B-F. In conclusion, at least for this application, PSS greatly improves the prediction accuracy that fits the interests of modeling noted earlier and, at the same time, provides efficient fast algorithm compared to the E-M based E-Lasso and regression based B-F procedure.

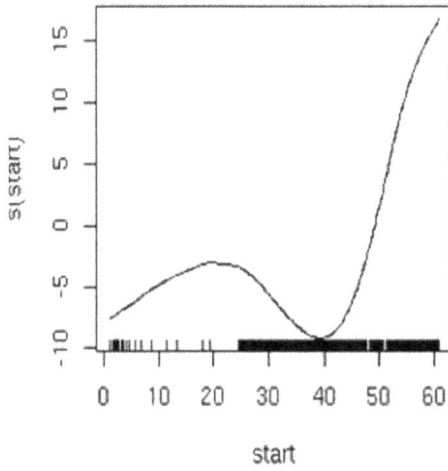

Fig. 7.5   *Fitted Smoothing Spline: Transfer Duration vs Start Time.*

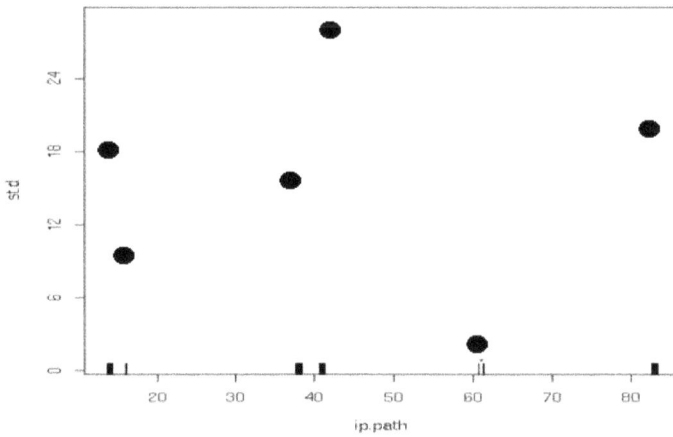

Fig. 7.6   *Estimated Non-zero Standard Deviation.*

7.6   Exercises

7.1. Show that the Gaussian linear model (see the beginning of Section 7.2) is a special case of GLM with the link function being the identity function, $g(\mu) = \mu$.

7.2. Consider the following simple example with $n = 1$. Let (x, y) be the observation. Compare the penalized least squares (PLS) that finds β to minimize $(y - \beta x)^2 + \lambda \beta^2$ and the PLS that finds β to minimize $(y - \beta x)^2 + \lambda |\beta|$, where in each case λ is a regularization parameter. Show that, for PLS with the L^2 penalty, the PLS estimator of β is not zero no matter how large λ is; on the other hand, for the PLS with the L^1 penalty, the PLS estimator of β is exactly zero if λ is sufficiently large.

7.3. For the case of regression variable selection, verify the claim (either by theoretical derivation, or via a numerical example) that if M_1 is a submodel of M_2, then $\lambda(M_1) \geq \lambda(M_2)$, where $\lambda(M)$ is defined in Section 7.3.

7.4. Show that, by treating the ξ_i's in Section 7.4 as observed, the complete-data log-likelihood can be expressed as (7.10).

Chapter 8

Model Selection with Incomplete Data

8.1 Introduction

The missing-data problem has a long history [e.g., Afifi and Elashoff (1966) 1966, Hartley and Hocking (1971)]. While there is an extensive literature on statistical analysis with missing or incomplete data [e.g., Rubin (1976), Dempster *et al.* (1977), Robins *et al.* (1995), Rotnitzky *et al.* (1998), Little and Rubin (2002)], the literature on model selection in the presence of missing data is relatively sparse. Existing model selection procedures face special challenges when confronted with missing or incomplete data. Obviously, the naive complete-data-only strategy is inefficient, sometimes even unacceptable by the practitioners due to the overwhelmingly wasted information. For example, in a study of backcross experiments [e.g., Lander and Botstein (1989), Zeng (1993), Jansen (1993), Broman and Speed (2002)], a data set was obtained by researchers at UC-Riverside [personal communications; see Zhan *et al.* (2011) for a related work]. Out of the 150 or so subjects, only 4 have complete data records. Situations like this are, unfortunately, the reality that we often have to deal with, and the main subject of this chapter.

Fuchs (1982) proposed to use the E-M algorithm [Dempster *et al.* (1977)] for the ML estimation under a log-linear model with missing data, and then test for goodness-of-fit based on the ML estimation in order to choose an appropriate model. Motivated by the predictive divergence for incomplete observation models [PDIO; Shimodaira (1994)], Cavanaugh and Shumway (1998) derived an AIC for model selection in the presence of incomplete data. A similar approach was considered by Seghouane *et al.* (2005), in which the authors obtained an unbiased estimator of the complete-data Kullback-Leibler symmetric divergence. Bueso *et al.* (1999) used the E-M algorithm to compute the minimum description length [MDL; Rissanen (1983)] for model selection, when only incomplete data are available. Sebastiani and Ramoni (2001) discussed a Bayesian approach for the selection of decomposable models by maximizing the posterior probability of a candidate model, and showed how to do this with incomplete data. Hens *et al.* (2006) considered a modification of the AIC based on reweighting incomplete and design-based samples. Claeskens and Consentino (2008) proposed some variations on the AIC based on the output of the E-M algorithm. The method is applicable to model selection problems with missing covariates, but the response variable is assumed to be fully observed. Schomaker *et al.* (2010) considered two approaches of handling the missing data in determining the weights in frequentist model averaging. The first is based on adjusting an existing criterion; while the second uses the unadjusted criterion but with the missing data replaced by their imputed values. Verbeke *et al.* (2008) offered a review of formal and informal model selection strategies with incomplete data, but the focus is on model comparison, instead of model selection. As noted by Ibrahim *et al.* (2008), while model comparisons "demonstrate the effect of assumptions on estimates and tests, they do not indicate which modeling strategy is best, nor do they specifically address model selection for a given class of models". The latter authors further proposed a class of model selection criteria based on the output of the E M algorithm. Also see Garcia *et al.* (2010). A potential drawback with the E-M approach of Ibrahim *et al.* (2008) is that the conditional expectation in the E-step is taken under the assumed (candidate) model, rather than an objective (true) model. Note that the complete-data log-likelihood is also based on the assumed model. Thus, by taking the conditional expection, again, under the assumed model, it may bring false supporting evidence for an incorrect model. The problem is sometimes referred to as "double-dipping", which we further illustrate in the next section.

8.2   A double-dipping problem

We use a simple example of logistic regression to illustrate the double-dipping problem and its impact on model selection.

Example 8.1. Suppose that one attempts to select a logistic model, $\text{logit}(p_i) = x_i'\beta$, where $p_i = \text{P}(Y_i = 1)$, $Y_1, \ldots, Y_n$ being independent, binary, observations, and x_i is a vector of covariates to be selected. Suppose that $y_1, \ldots, y_5$ are observed, and the rest of the y_i's are missing. Also, for simplicity, assume that all the x_i's are observed. The derivation below in this paragraph is based on MAR [Rubin (1976)] for simplicity. Let M_0 denote the intercept only model and suppose that the true model is not M_0. The complete-data log-likelihood under M_0 is given by

$$l = \sum_{i=1}^{n} \{y_i \log(p_0) + (1 - y_i) \log(1 - p_0)\}, \tag{8.1}$$

where $p_0 = e^{\beta_0}/(1 + e^{\beta_0})$ and β_0 is the intercept; thus, under M_0, we have

$$E(l|y_1, \ldots, y_5, \text{ all } x_i's)$$
$$= \sum_{i=1}^{5} \{y_i \log(p_0) + (1 - y_i) \log(1 - p_0)\}$$
$$+ (n - 5)\{p_0 \log(p_0) + (1 - p_0) \log(1 - p_0)\}. \tag{8.2}$$

If $y_i = 1, 1 \le i \le 5$, then, as $p_0 \to 1$, we have $E(l|y_1, \ldots, y_5, \text{ all } x_i's) \to 0$. On the other hand, under any other model, M, the corresponding log-likelihood is given by

$$l = \sum_{i=1}^{n} \{y_i \log(p_i) + (1 - y_i) \log(1 - p_i)\} \le 0,$$

hence $E(l|y_1, \ldots, y_5, \text{ all } x_i's) \le 0$, under M. This means that the maximized conditional expectation of l under M_0 (which is 0) is greater than or equal to the maximized conditional expectation of l under M (which is less than or equal to 0). Thus, the first term of any information criterion [see (1.5)] under M_0 is less than or equal to that under M. On the other hand, M_0 certainly has the smallest dimension. Therefore, M_0 will be selected as the optimal model by the IC criteria of Ibrahim *et al.* (2008), which, of course, is an incorrect model.

To further illustrate numerically, we carry out a simulation study under the following specific setting. Suppose that the candidate covariates include a continuous variable, x_1, whose values are generated from the standard normal distribution, and a binary indicator, x_2, whose values are generated from the Bernoulli(0.5) distribution. The following candidate models are considered: Model 0: $x_i'\beta = \beta_0$, Model j: $x_i'\beta = \beta_0 + \beta_j x_{ji}, j = 1, 2$, and Model 3: $x_i'\beta = \beta_0 + \beta_1 x_{1i} + \beta_2 x_{2i}$. Two scenarios are considered. In the first scenario, Model 1 is the true underlying model with the true parameters $\beta_0 = \beta_1 = 1$; in the second scenario, Model 3, which is the full model, is the true underlying model with the true parameters $\beta_0 = \beta_1 = 1, \beta_2 = -1$. Furthermore, the missing data indicators, M_i, which is 1 if y_i is missing, and

Table 8.1 **Empirical TP for Logistic Model Selection.**

Missing Data Mechanism	Sample Size	True Model = Model 1		True Model = Model 3	
		E-MS	IZT	E-MS	IZT
Case A	$n = 50$	0.787	0.483	0.213	0.136
	$n = 100$	0.965	0.738	0.467	0.216
Case B	$n = 50$	0.837	0.395	0.169	0.097
	$n = 100$	0.970	0.607	0.459	0.160

0 otherwise, are generated either under an *ignorable* mechanism, in which case $P(M_i = 1|y) = 0.5$ (case A), or under a *non-ignorable* mechanism, in which case $P(M_i = 1|y) = h(\psi_0 + \psi_1 y_i)$ with $h(x) = e^x/(1 + e^x)$ and the true parameters $\psi_0 = 0.5$ and $\psi_1 = 0.2$ (case B). See Section 8.7 for more details. We apply the method of Ibrahim *et al.* (2008) with the BIC penalty (see Section 1.1), denoted by IZT, under two different sample sizes, $n = 50$ and $n = 100$. A comparing method, called E-MS (to be introduced in Section 8.4), here in conjunction with the BIC, is also applied to the same simulated data. Results of the empirical true-positive (TP), based on 1,000 simulations, are reported in Table 8.1. It is seen that IZT performs considerably worse than E-MS under all scenarios, cases, and sample sizes. Note that both methods perform much worse under Model 3 than under Model 1, which is not surprising–the BIC is known to over-penalize "larger" models, especially the full model [e.g., Jiang *et al.* (2008)]. Furthermore, the performance of E-MS does not seem to be affected by the different missing data mechanisms (see Section 8.7 for more discussion), while IZT appears to perform worse under the non-ignorable missing data setting (case B).

In the following sections, we discuss two alternative approaches to model selection with incomplete data. Both approaches are associated with the idea of the E-M algorithm [Dempster *et al.* (1977)]. A common feature of these approaches is that they do not suffer from the double-dipping problem noted above. The first approach, called EMAF algorithm, works naturally with the AF idea (see Section 3.1). The second approach, known as the E-MS algorithm, works more generally with essentially any model selection procedure that is built for the complete-data situations, including the information criteria and various fence methods.

8.3 The EMAF algorithm

Classical model selection assumes that there is a true model among the class of candidate models. While this assumption may not always be practical, it is the basis for consistent model selection. For the simplicity of illustration, let us assume, for now, that this classical assumption holds. Furthermore, we assume that the candidate models include a full model. It then follows that the full model is, at least, a correct model, even though it may not be the most efficient one. Thus, in the presence of missing data, we can run the E-M to obtain the MLE of the parameters, under the full model. Note that, here, we do not have the double-dipping problem (see the previous section), because the conditional expectation (under the full model) is "objective". Once the parameter estimates are obtained, we can use model-based (or parametric) bootstrap to draw samples, under the full model, as in the AF (see Section 3.1). The best part of this strategy is that, when one draws the model-based bootstrap samples, one draws samples of the complete data, rather than data with the missing values. Therefore, one can apply any existing model selection procedure that is built for the complete-data situation, such as the information criteria and fence, to the bootstrap sample. For example, suppose that one intends to use the BIC for model selection. Then, for each bootstrap sample of the complete data, the BIC is applied, and a model is selected as the optimal model. Suppose that B bootstrap samples are drawn. The model with the highest frequency of being selected as the optimal model, among the bootstrap samples, is the (final) optimal model. Again, the latter idea of choosing the optimal model is similar to AF. For these reasons, we call this procedure EMAF algorithm. Again, the rationale is the following. Consider, for example, regression variable selection. The true variables are those corresponding to the nonzero regression coefficients under the full model. Although the MLEs of the zero coefficients are not exactly zero, as argued in Jiang *et al.* (2008), they are close to zero, or at least much smaller in absolute values compared to those of the nonzero coefficients (provided that the variables are suitably standardized so that they are of the same scale). Thus, by bootstrapping under the full model, the AF allows the data to "speak" on which subset of variables they most support, based on available information.

The EMAF is particularly suitable with the AF (see Section 3.2), or the IF (see Section 5.1). Also note a similarity between EMAF and the observed best prediction [OBP; Jiang *et al.* (2011a)] in the sense of entertaining two models; one is an assumed, more restrictive model and one is a broader model, which is (much) more likely to be a true model. The assumed model is more of interest in that it utilizes the explicit relationship between variables whose data are available; however, the evaluation of a statistical performance measure, here the empirical probability, is based on an objective model that is correct, or close to be correct, even though the objective model is not as useful as the more restrictive model. This way, we avoid falling into the trap of double-using the assumed model, which leads to the double-dipping.

Building on the EMAF idea, we can extend the method to situations where a true model may not exist, or exists but not among the candidates, as long as one can find a broader model that is approximately correct, and bootstrap under this model. We illustrate the implementation of EMAF using a specific example. More examples are considered in Section 8.5.

Example 8.2. (QTL mapping in backcross experiments) Quantitative trait loci (QTL) mapping in genetics has been extensively studied (e.g., Lander and Botstein (1989), Zeng (1993), Jansen (1993)). More recently, Broman and Speed (2002) modified the BIC and applied it to QTL mapping in backcross experiments. The method is for complete-data analysis only. In practice, however, missing data are often present. For example, as mentioned earlier, in the backcross data set obtained by the researchers at UC-Riverside, less than 3% of the data have the complete records, i.e., without the missing values.

Following Broman and Speed (2002), we have a conditional linear regression model for the phenotype variable, y, such that, given the marker indicators, x, we have $y_i = \sum_{k=1}^{r} \sum_{j \in M_k} \beta_{jk} x_{ijk} + \epsilon_i$, where r is the number of chromosomes, M_k is a subset of $\{1, \ldots, q\}$ and q is the number of markers on each chromosome, and ϵ_i is a normal error, with mean zero and unknown variance σ^2. The ϵ_i's are uncorrelated and also independent with the x_{ijk}'s. Furthermore, the marker indicators, x_{ijk}, are assumed to be a Markov chain within each chromosome with $P(x_{i1k} = 0) = P(x_{i1k} = 1) = 1/2$ (Mendel's rule) and $P(x_{ij+1k} = 1 | x_{ijk} = 0) = P(x_{ij+1k} = 0 | x_{ijk} = 1) = \theta$, where θ is the (unknown) *recombination fraction*. The problem of interest is to identify the subset $M = (M_1, \ldots, M_r)$, which is viewed as a model selection problem as in Broman and Speed (2002). Because the distribution of the x's is not model-dependent, the model selection is based on the conditional log-likelihood of y given x. The complete data log-likelihood, under the full model, can be expressed as $l = l_{y|x} + l_x$, where

$$l_{y|x} = c_1 - \frac{1}{2} \left\{ n \log \sigma^2 + \sigma^{-2} \sum_{i=1}^{n} (y_i - x'_{f,i} \beta_f)^2 \right\}, \tag{8.3}$$

$$l_x = c_2 + nr(q-1) \log(1-\theta) + \{\log \theta - \log(1-\theta)\} \sum_{i=1}^{n} s(\mathcal{X}_i) \tag{8.4}$$

with $s(\mathcal{X}_i) = \sum_{k=1}^{r}\sum_{j=1}^{q-1}(x_{ijk} + x_{ij+1k} - 2x_{ijk}x_{ij+1k}) = \#$ of cases among $x_{ijk}, 1 \leq k \leq r, 1 \leq j \leq q-1$ such that $|x_{ijk} - x_{ij+1k}| = 1$, where c_1, c_2 are constants (Exercise 8.2). Note that θ is involved only in l_x, hence, the MLE of θ involves only the x data. Thus, we first update the estimate of θ based on the x data. Let $\hat{\theta}_c$ denote the current estimate. Then, we have

$$E_c(l_x|\mathcal{X}_o) \propto nr(q-1)\log(1-\theta)$$

$$+ \{\log\theta - \log(1-\theta)\}\sum_{i=1}^{n} E_c\{s(\mathcal{X}_i)|\mathcal{X}_{o,i}\} \qquad (8.5)$$

(Exercise 8.3), where and hereafter E_c means (conditional) expectation under the current estimates, and $\mathcal{X}_{o,i}$ denotes all the observed x's among $x_{ijk}, 1 \leq j \leq q, 1 \leq k \leq r$. Similarly, let $\mathcal{X}_{m,i}$ denote all the missing x's among $x_{ijk}, 1 \leq j \leq q, 1 \leq k \leq r$. Let f_c denotes the (conditional) pmf, or pdf, under the current estimates. It can be shown that

$$f_c(\mathcal{X}_i) = \frac{(1-\hat{\theta}_c)^{r(q-1)}}{2^r}\left(\frac{\hat{\theta}_c}{1-\hat{\theta}_c}\right)^{s(\mathcal{X}_i)}, \qquad (8.6)$$

where $\mathcal{X}_i = (x_{ijk})_{1\leq j\leq q, 1\leq k\leq r}$. Thus, we have

$$f_c(\mathcal{X}_{m,i}|\mathcal{X}_{o,i}) = \frac{f_c(\mathcal{X}_i)}{\sum_{\mathcal{X}_{m,i}} f_c(\mathcal{X}_i)} = \frac{\hat{\gamma}_c^{s(\mathcal{X}_i)}}{\sum_{\mathcal{X}_{m,i}} \hat{\gamma}_c^{s(\mathcal{X}_i)}},$$

where $\hat{\gamma}_c = \hat{\theta}_c/(1-\hat{\theta}_c)$, and $\sum_{\mathcal{X}_{m,i}}$ is over all the missing x's among $x_{ijk}, 1 \leq j \leq q, 1 \leq k \leq r$ (each taking the value of 0 or 1). It follows that

$$E_c\{s(\mathcal{X}_i)|\mathcal{X}_{o,i}\} = \frac{\sum_{\mathcal{X}_{m,i}} s(\mathcal{X}_i)\hat{\gamma}_c^{s(\mathcal{X}_i)}}{\sum_{\mathcal{X}_{m,i}} \hat{\gamma}_c^{s(\mathcal{X}_i)}}. \qquad (8.7)$$

By (8.5) and (8.7), it is easy to obtain the maximizer of (8.5), that is, the update for the θ estimate:

$$\hat{\theta} = \frac{1}{nr(q-1)}\sum_{i=1}^{n}\frac{\sum_{\mathcal{X}_{m,i}} s(\mathcal{X}_i)\hat{\gamma}_c^{s(\mathcal{X}_i)}}{\sum_{\mathcal{X}_{m,i}} \hat{\gamma}_c^{s(\mathcal{X}_i)}}. \qquad (8.8)$$

As a comparison, note that the MLE of θ based on all of the x's is

$$\hat{\theta}_a = \frac{\sum_{i=1}^{n} s(\mathcal{X}_i)}{nr(q-1)} \qquad (8.9)$$

(Exercise 8.4), where the subscript a refers to "all of the x's".

We next update the estimates of β_f and σ^2. Again, let $\hat{\beta}_{f,c}, \hat{\sigma}_c^2$ denote the current estimates. Then, we have

$$
\mathrm{E}_c(l_{y|x}|y_o, \mathcal{X}_o)
$$

$$
\propto -\frac{1}{2}\left[n \log \sigma^2 + \frac{1}{\sigma^2}\sum_{i=1}^{n}\mathrm{E}_c\left\{(y_i - x'_{f,i}\beta_f)^2 \,\big|\, y_{o,i}, \mathcal{X}_{o,i}\right\}\right], \quad (8.10)
$$

where $y_{o,i} = y_i$ if y_i is observed, and the notation $y_{o,i}$ disappears if y_i is missing. Similarly, $y_{m,i} = y_i$ if y_i is missing, and the notation $y_{m,i}$ disappears if y_i is observed. To evaluate the conditional expectations in (8.10), we need to obtain $f_c(y_{m,i}, \mathcal{X}_{m,i}|y_{o,i}, \mathcal{X}_{o,i})$. Note that

$$
f_c(y_{m,i}, \mathcal{X}_{m,i}|y_{o,i}, \mathcal{X}_{o,i}) = \frac{f_c(y_i, \mathcal{X}_i)}{f_c(y_{o,i}, \mathcal{X}_{o,i})} = \frac{f_c(y_i|\mathcal{X}_i)f_c(\mathcal{X}_i)}{f_c(y_{o,i}, \mathcal{X}_{o,i})}, \quad (8.11)
$$

where $\mathcal{X}_i = (x_{ijk})_{1\le j\le q, 1\le k\le r}$, and

$$
f_c(y_i|\mathcal{X}_i) = \frac{1}{\sqrt{2\pi\hat{\sigma}_c^2}}\exp\left\{-\frac{(y_i - x'_{f,i}\hat{\beta}_{f,c})^2}{2\hat{\sigma}_c^2}\right\}. \quad (8.12)
$$

If y_i is observed, then,

$$
f_c(y_{o,i}, \mathcal{X}_{o,i}) = \sum_{\mathcal{X}_{m,i}}f_c(y_i, \mathcal{X}_i) = \sum_{\mathcal{X}_{m,i}}f_c(y_i|\mathcal{X}_i)f_c(\mathcal{X}_i).
$$

Thus, by (8.11), (8.12) and (8.6), we have

$$
f_c(y_{m,i}, \mathcal{X}_{m,i}|y_{o,i}, \mathcal{X}_{o,i}) = \frac{f_c(y_i|\mathcal{X}_i)f_c(\mathcal{X}_i)}{\sum_{\mathcal{X}_{m,i}}f_c(y_i|\mathcal{X}_i)f_c(\mathcal{X}_i)}
$$

$$
= \frac{\exp\{-(2\hat{\sigma}_c^2)^{-1}(y_i - x'_{f,i}\hat{\beta}_{f,c})^2\}\hat{\gamma}^{s(\mathcal{X}_i)}}{\sum_{\mathcal{X}_{m,i}}\exp\{-(2\hat{\sigma}_c^2)^{-1}(y_i - x'_{f,i}\hat{\beta}_{f,c})^2\}\hat{\gamma}^{s(\mathcal{X}_i)}},
$$

where $\hat{\gamma} = \hat{\theta}/(1 - \hat{\theta})$ and $\hat{\theta}$ is given by (8.8). It follows that (Exercise 8.5)

$$
\mathrm{E}_c\left\{(y_i - x'_{f,i}\beta_f)^2|y_{o,i}, \mathcal{X}_{o,i}\right\}
$$

$$
= \frac{\sum_{\mathcal{X}_{m,i}}(y_i - x'_{f,i}\beta_f)^2\exp\{-(2\hat{\sigma}_c^2)^{-1}(y_i - x'_{f,i}\hat{\beta}_{f,c})^2\}\hat{\gamma}^{s(\mathcal{X}_i)}}{\sum_{\mathcal{X}_{m,i}}\exp\{-(2\hat{\sigma}_c^2)^{-1}(y_i - x'_{f,i}\hat{\beta}_{f,c})^2\}\hat{\gamma}^{s(\mathcal{X}_i)}}. \quad (8.13)
$$

If y_i is missing, then, we have

$$
f_c(y_{o,i}, \mathcal{X}_{o,i}) = f_c(\mathcal{X}_{o,i}) = \sum_{\mathcal{X}_{m,i}}f_c(\mathcal{X}_i);
$$

hence, by (8.11) and (8.6), we have

$$
f_c(y_{m,i}, \mathcal{X}_{m,i}|y_{o,i}, \mathcal{X}_{o,i}) = \frac{f_c(y_i|\mathcal{X}_i)f_c(\mathcal{X}_i)}{\sum_{\mathcal{X}_{m,i}}f_c(\mathcal{X}_i)} = \frac{f_c(y_i|\mathcal{X}_i)\hat{\gamma}^{s(\mathcal{X}_i)}}{\sum_{\mathcal{X}_{m,i}}\hat{\gamma}^{s(\mathcal{X}_i)}}.
$$

By (8.12), we have

$$\int (y_i - x'_{f,i}\beta_f)^2 f_c(y_i|\mathcal{X}_i)dy_i = \hat{\sigma}_c^2 + \{x'_{f,i}(\hat{\beta}_{f,c} - \beta_f)\}^2.$$

Thus, we obtain (Exercise 8.5)

$$E_c\left\{(y_i - x'_{f,i}\beta_f)^2|y_{o,i}, \mathcal{X}_{o,i}\right\}$$
$$= \hat{\sigma}_c^2 + \frac{\sum_{\mathcal{X}_{m,i}}\{x'_{f,i}(\hat{\beta}_{f,c} - \beta_f)\}^2\hat{\gamma}^{s(\mathcal{X}_i)}}{\sum_{\mathcal{X}_{m,i}}\hat{\gamma}^{s(\mathcal{X}_i)}}. \quad (8.14)$$

(8.13) and (8.14) may look a little complicated, but, the bottom line is: they are quadratic functions of β_f. In fact, if we define

$$w_{o,i}(y_i, \mathcal{X}_i) = \frac{\exp\{-(2\hat{\sigma}_c^2)^{-1}(y_i - x'_{f,i}\hat{\beta}_{f,c})^2\}\hat{\gamma}^{s(\mathcal{X}_i)}}{\sum_{\mathcal{X}_{m,i}}\exp\{-(2\hat{\sigma}_c^2)^{-1}(y_i - x'_{f,i}\hat{\beta}_{f,c})^2\}\hat{\gamma}^{s(\mathcal{X}_i)}},$$

and $w_{m,i}(\mathcal{X}_i) = \hat{\gamma}^{s(\mathcal{X}_i)}/\sum_{\mathcal{X}_{m,i}}\hat{\gamma}^{s(\mathcal{X}_i)}$, and $n_m = |I_m|$ (cardinality), then, by letting the derivatives of (8.10) equal to zero, we find the following solution:

$$\hat{\beta}_f = \left\{\sum_{i\in I_o}\sum_{\mathcal{X}_{m,i}} w_{o,i}(y_i, \mathcal{X}_i)x_{f,i}x'_{f,i} + \sum_{i\in I_m}\sum_{\mathcal{X}_{m,i}} w_{m,i}(\mathcal{X}_i)x_{f,i}x'_{f,i}\right\}^{-1}$$

$$\times \left[\sum_{i\in I_o}\sum_{\mathcal{X}_{m,i}} w_{o,i}(y_i, \mathcal{X}_i)x_{f,i}y_i\right.$$

$$\left. + \left\{\sum_{i\in I_m}\sum_{\mathcal{X}_{m,i}} w_{m,i}(\mathcal{X}_i)x_{f,i}x'_{f,i}\right\}\hat{\beta}_{f,c}\right], \quad (8.15)$$

$$\hat{\sigma}^2 = \left(\frac{n_m}{n}\right)\hat{\sigma}_c^2 + \frac{1}{n}\left[\sum_{i\in I_o}\sum_{\mathcal{X}_{m,i}} w_{o,i}(y_i, \mathcal{X}_i)(y_i - x'_{f,i}\hat{\beta}_f)^2\right.$$

$$\left. + \sum_{i\in I_m}\sum_{\mathcal{X}_{m,i}} w_{m,i}(\mathcal{X}_i)\{x'_{f,i}(\hat{\beta}_{f,c} - \hat{\beta}_f)\}^2\right]. \quad (8.16)$$

A reasonable initial estimator for θ is $\hat{\theta}_0 =$ proportion of observed cases in which x_{ijk} and x_{ij+1k} are different. As for the initial estimator of β_f, the vector of regression coefficients under the full model, note that the idea of least squares (LS) fit to the regression is to find the parameter estimates that minimizes $\sum_{i=1}^n\{y_i - E_f(y_i|\mathcal{X})\}^2$, where E_f denotes expectation under the full model and $\mathcal{X}$ the complete x's. Due to the missing data, it is natural to replace this by $\sum_{i\in I_o}\{y_i - E_f(y_i|\mathcal{X}_o)\}^2$, where I_o denotes the subset of indexes i so that y_i is observed, and $\mathcal{X}_o$ the observed x's. It follows that $\mathrm{E}_f(y_i|\mathcal{X}_o) = \sum_{k=1}^{r} \sum_{j=1}^{q} \beta_{jk} \mathrm{E}_f(x_{ijk}|\mathcal{X}_{o,i})$, under the assumed (full) model, where $\mathcal{X}_{o,i}$ denotes the observe x's for the ith subject. We have $\mathrm{E}_f(x_{ijk}|\mathcal{X}_{o,i}) = x_{ijk}$ if the latter is observed, and an expression of the conditional expectation can be easily obtained under the Markov chain model, with θ replaced by $\hat{\theta}_0$, if x_{ijk} is missing. We then fit the LS problem with $y_i, i \in I_o$ as the responses and $\mathrm{E}_f(x_{ijk}|\mathcal{X}_{o,i})$'s, $i \in I_o$, as the predictors, to obtain the initial estimator $\hat{\beta}_{f,0}$, for β_f. The initial estimator for σ^2, $\hat{\sigma}_0^2$, is the RSS of this LS fit divided by $|I_o| - qr$.

Let $\hat{\theta}, \hat{\beta}_f, \hat{\sigma}^2$ denote the MLEs of $\theta, \beta_f, \sigma^2$, respectively, obtained via the E-M algorithm, then it is straightforward to draw bootstrap samples of the complete data. Namely, first draw X_f^* using the Markov chain with $\theta = \hat{\theta}$, where X_f denote X under the full model; then let $y^* = X_f^* \hat{\beta}_f + \hat{\sigma}\xi^*$, where the components of ξ^* are independently drawn from $N(0,1)$.

8.4   The E-MS algorithm

The E-M algorithm is well known for parameter estimation in the presence of missing data. On the other hand, model selection, as another component of model identification, may also be viewed as parameter estimation, with the parameter being [the identification (ID) number of] the model and the parameter space being the (ID numbers of the) model space. Namely, we combine the parameters with the model under which the parameters are defined. So, at the current stage of the iteration, we have the current model, M_c, as well as the current estimates of the parameters, $\hat{\theta}_c$, under M_c. Let $Q(M) = Q(Y, M, \theta_M)$ be a measure of lack-of-fit, where Y represents the complete data, M a candidate model, and θ_M the vector of parameters under M. We take the conditional expectation of $Q(M)$ under M_c, with the parameters under M_c, $\theta_{M_c} \equiv \theta_c$ being $\hat{\theta}_c$, given the observed data, Y_o. Denote the conditional expectation by $\mathrm{E}_c\{Q(M)|Y_o\}$. This is the E-step.

In the next step, we carry out model selection using $\mathrm{E}_c\{Q(M)|Y_o\}$ as the measure of lack-of-fit. To do so, we first find

$$Q_c(M) = \inf_{\theta_M \in \Theta_M} \mathrm{E}_c\{Q(M)|Y_o\},$$

where Θ_M is the parameter space under M. We can use Q_c in a generalized information criterion (GIC; e.g., Nishii 1984, Shibata 1984) setting, in which the optimal model, M_{opt}, is found by minimizing $Q_c(M) + \lambda_n|M|$ over $M \in \mathcal{M}$, the class of candidate models, where λ_n is a penalty that depends on the sample size, n, and $|M|$ is the dimension of M. Alternatively, we may use the fence methods based on Q_c, for example, (3.2) with Q replaced by, Q_c. This is the MS-step, where MS stands for "model selection". We then replace M_c by M_{opt}, found in the MS-step, and $\hat{\theta}_c$ by $\hat{\theta}_{opt}$, where $\hat{\theta}_{opt}$ is the parameter vector under M_{opt} corresponding to the minimizer of $E_c\{Q(M_{opt})|Y_o\}$ over $\theta_{M_{opt}} \in \Theta_{M_{opt}}$, and return to the E-step.

The E-MS is particularly simple when working repeatedly under the EMAF framework. Take $M_0 = M_f$ as the initial model. Run the EMAF to obtain the optimal model, denoted by M_1. Run the EAMF again with M_0 replaced by M_1. This means that we do the same thing as EMAF, described in the previous section, except with the roles of M_f replaced by M_1. This leads to our next optimal model, M_2. Run the EMAF again with M_1 replaced by M_2, and so on. We illustrate with an example.

Example 8.3. (Linear regression) The classical linear regression is a conditional model, in which the distribution of the covariates (or predictors) is not specified. As is well known, such a model may not directly work with the E-M algorithm, if some of the covariates are also missing. Little and Rubin (2002) proposed the following model for the joint distribution of the response and predictors in a linear regression. Suppose that the candidate predictors can be listed as $x_1, \ldots, x_p, x_{p+1}, \ldots, x_{p+q}$ such that $x_1, \ldots, x_p$ are continuous and $x_{p+1}, \ldots, x_{p+q}$ are discrete, or categorical, predictors (in case there is an intercept, the corresponding constant, 1, is considered as the first discrete/categorical predictor). Furthermore, let $v_1, \ldots, v_s$ be all the possible (vector-valued) values for $x_d = (x_{p+1}, \ldots, x_{p+q})'$. Let $x_{i,d}$ be the x_d corresponding to the ith observation, and $x_{i,c}$ be the vector $(x_1, \ldots, x_p)'$ corresponding to the ith observation, and $x_i = (x'_{i,c}, x'_{i,d})'$. The assumptions are:

(i) $y_i, x_i, i = 1, \ldots, n$ are independent.

(ii) For each i, $x_{i,d}$ has the probability distribution $P(x_{i,d} = v_r) = \pi_r, 1 \leq r \leq s$, where the π_r's are unknown probabilities such that $\sum_{r=1}^{s} \pi_r = 1$.

(iii) Given $x_{i,d} = v_r$, $x_{i,c}$ has a multivariate normal distribution with mean μ_r and covariance matrix Ω, where $\mu_r, 1 \leq r \leq s$ are unknown vectors, and Ω is an unknown covariance matrix that does not depend on r.

(iv) Given x_i, y_i is normal with mean $x'_i\beta$ and variance σ^2, where β is an unknown $(p+q)$-dimensional vector of regression coefficients, and σ^2 is an unknown variance.

The assumptions (i)–(iii) are for the full model. More generally, we are interested in a model, M, for the conditional distribution (iv). Write

$x_{i,M} = (x'_{i,M,c}, x'_{i,M,d})'$, and $\beta_M = (\beta'_{M,c}, \beta'_{M,d})'$. Then, under M, (iv) is replaced by

(iv-M) Given x_i, $y_i \sim N(x'_{i,M}\beta_M, \sigma^2)$,

while parts (i)–(iii) of the model are unchanged.

Let y, x, x_c, x_d denote the data for the $y_i, x_i, x_{i,c}, x_{i,d}$, respectively, across $1 \le i \le n$. Then, it can be shown that the complete-data log-likelihood has the expression

$$l = c - \frac{n}{2}(\log \sigma^2 + \log|\Omega|) + \sum_{r=1}^{s} n_r \log \pi_r - \frac{1}{2}\sum_{r=1}^{s}\sum_{i=1}^{n}$$
$$1_{(x_{i,d}=v_r)}(x_{i,c} - \mu_r)'\Omega^{-1}(x_{i,c} - \mu_r)$$
$$- \frac{1}{2\sigma^2}\sum_{i=1}^{n}(y_i - x'_{i,M}\beta_M)^2, \tag{8.17}$$

where c is a constant. Note that the maximum likelihood is a constrained maximization problem, namely, $\max l$ subject to $\sum_{r=1}^{s} \pi_r = 1$. Define $\mathcal{L} = l + \lambda(\sum_{r=1}^{s} \pi_r - 1)$. Then, the MLE of the parameters, plus the Lagrange multiplier λ, is a stationary point of $\mathcal{L}$.

Before considering E-MS, let us first look at EMAF as this sets out the basic element for E-MS with AF (Chapter 3). Let M_f denote the full model and β_f the β under M_f. Let E_f denote the conditional expectation under M_f and the current estimates of parameters, under M_f, including β_f, σ^2, $\mu_r, \pi_r, 1 \le r \le s$, and Ω. Let y_o, x_o denote the observed y, x, respectively. Then, by (8.17), with $M = M_f$, we have

$$\bar{\mathcal{L}}_f \equiv E_f(\mathcal{L}|y_o, x_o)$$
$$= \cdots - \frac{n}{2}\log\sigma^2 - \frac{1}{2\sigma^2}\sum_{i=1}^{n}E_f\left\{(y_i - x'_{i,f}\beta_f)^2|y_o, x_o\right\}, \tag{8.18}$$

where $\cdots$ does not depend on β_f and σ^2. From (8.18), we obtain the updates for β_f and σ^2,

$$\hat{\beta}_f = S_2^{-1}S_1,$$
$$\hat{\sigma}^2 = \frac{S_0 - S_1'S_2^{-1}S_1}{n} \tag{8.19}$$

with $S_0 = \sum_{i=1}^{n}E_f(y_i^2|y_o, x_o)$, $S_1 = \sum_{i=1}^{n}E_f(x_{i,f}y_i|y_o, x_o)$, $S_2 = $

$\sum_{i=1}^{n} \mathrm{E}_{\mathrm{f}}(x_{i,\mathrm{f}} x'_{i,\mathrm{f}} | y_{\mathrm{o}}, x_{\mathrm{o}})$. Furthermore, we have (Exercise 8.6)

$$\hat{\mu}_r = \frac{\sum_{i=1}^{n} \mathrm{E}_{\mathrm{f}}\{1_{(x_{i,\mathrm{d}}=v_r)} x_{i,\mathrm{c}} | y_{\mathrm{o}}, x_{\mathrm{o}}\}}{\sum_{i=1}^{n} \mathrm{P}_{\mathrm{f}}(x_{i,\mathrm{d}} = v_r | y_{\mathrm{o}}, x_{\mathrm{o}})},$$

$$\hat{\pi}_r = \frac{\mathrm{E}_{\mathrm{f}}(n_r | y_{\mathrm{o}}, x_{\mathrm{o}})}{\sum_{t=1}^{s} \mathrm{E}_{\mathrm{f}}(n_t | y_{\mathrm{o}}, x_{\mathrm{o}})}, \quad 1 \le r \le s,$$

$$\hat{\Omega} = \frac{1}{n} \sum_{r=1}^{s} \sum_{i=1}^{n} \mathrm{E}_{\mathrm{f}}\{1_{(x_{i,\mathrm{d}}=v_r)}(x_{i,\mathrm{c}} - \hat{\mu}_r)(x_{i,\mathrm{c}} - \hat{\mu}_r)' | y_{\mathrm{o}}, x_{\mathrm{o}}\}. \quad (8.20)$$

It remains to evaluate the conditional expectations involved in (8.19), (8.20). Let y_{m}, x_{m}, $x_{\mathrm{c,m}}$, and $x_{\mathrm{d,m}}$ denote the missing parts of y, x, x_{c}, and x_{d}, respectively. Although it is possible to obtain the conditional density, $f_M(y_{\mathrm{m}}, x_{\mathrm{m}} | y_{\mathrm{o}}, x_{\mathrm{o}})$, the result is not a standard distribution (e.g., normal), under which the conditional expectations can be easily obtained analytically. Alternatively, one may consider sampling from the conditional distribution, and use the Monte-Carlo method to compute the conditional expectations. To do so, first note that it is easy to show that one can sample from the joint conditional distribution by sampling independently from the conditional distribution for each subject (Exercise 8.7). To sample from the subject conditional distribution, note that $f_{M,i}(y_{i,\mathrm{m}}, x_{i,\mathrm{m}} | y_{i,\mathrm{o}}, x_{i,\mathrm{o}}) \propto f_{M,i}(y_i, x_i) \propto$

$$\exp\left[\sum_{r=1}^{s} 1_{(x_{i,\mathrm{d}}=v_r)} \left\{ \log \pi_r - \frac{1}{2}(x_{i,\mathrm{c}} - \mu_r)' \Omega^{-1}(x_{i,\mathrm{c}} - \mu_r) \right\} \right.$$
$$\left. - \frac{(y_i - x'_{i,M} \beta_M)^2}{2\sigma^2} \right], \quad (8.21)$$

where $\propto$ means that the expression is up to a function of $y_{i,\mathrm{o}}, x_{i,\mathrm{o}}$, which is considered constant during the sampling of $y_{i,\mathrm{m}}, x_{i,\mathrm{m}}$. Next, we employ the Metropolized independence sampler (MIS, e.g., Liu (2004), p. 115), which is a special case of the Metropolis-Hastings algorithm, as follows. Write $z = (y_{i,\mathrm{m}}, x_{i,\mathrm{m}})$, and $f(z) =$ the right side of (8.21) (note that $y_{i,\mathrm{o}}, x_{i,\mathrm{o}}$ are held fix in y_i, x_i). Given the current state $z^{(t)}$, (a) draw $z \sim g(z)$, where $g(\cdot)$ is a *trial density* whose expression is known, up to a constant, and from which one knows how to sample; (b) simulate $u \sim \mathrm{Uniform}[0, 1]$ and let

$$z^{(t+1)} = \begin{cases} z, & \text{if } u \le \min\{1, w(z)/w(z^{(t)})\}, \\ z^{(t)}, & \text{otherwise,} \end{cases}$$

where $w(z) = f(z)/g(z)$ is the *importance sampling weight*. The algorithm generates a Markov chain that converges (in distribution) to its stationary distribution, which is the target distribution on the left side of (8.21).

It remains to choose the trial density g. Let $z = (x_{i,\mathrm{d,m}}, x_{i,\mathrm{c,m}}, y_{i,\mathrm{m}})$, where, for the moment, assume that all three components of z are non-empty. We proceed as follows:

(I) First draw $x_{i,\mathrm{d,m}}$ from $f(x_{i,\mathrm{d,m}}|x_{i,\mathrm{d,o}})$. Note that, given the missing value pattern for x_i, each vector v_r is partitioned as $v_{r,i,\mathrm{o}}$ and $v_{r,i,\mathrm{m}}$, with the notations understood in obvious ways. Then, given $x_{i,\mathrm{d,o}} = v_{r,i,\mathrm{o}}$, the possible values of $x_{i,\mathrm{d,m}}$ are $v_{\tilde{r},i,\mathrm{m}}, 1 \leq \tilde{r} \leq s$ such that $v_{\tilde{r},i,\mathrm{o}} = v_{r,i,\mathrm{o}}$. In other words, define $R(v) = \{1 \leq \tilde{r} \leq s : v_{\tilde{r},i,\mathrm{o}} = v\}$. Then, the possible values of $x_{i,\mathrm{d,m}}$ are $v_{\tilde{r},i,\mathrm{m}}, \tilde{r} \in R(v_{r,i,\mathrm{o}})$; and, for any $\tilde{r} \in R(v_{r,i,\mathrm{o}})$, we have

$$\mathrm{P}(x_{i,\mathrm{d,m}} = v_{\tilde{r},i,\mathrm{m}}|x_{i,\mathrm{d,o}} = v_{r,i,\mathrm{o}}) = \frac{\mathrm{P}(x_{i,\mathrm{d,m}} = v_{\tilde{r},i,\mathrm{m}}, x_{i,\mathrm{d,o}} = v_{\tilde{r},i,\mathrm{o}})}{\mathrm{P}(x_{i,\mathrm{d,o}} = v_{r,i,\mathrm{o}})}$$

$$= c\mathrm{P}(x_{i,\mathrm{d}} = v_{\tilde{r}}),$$

where c does not depend on $\tilde{r}$. By summing over $\tilde{r} \in R(v_{r,i,\mathrm{o}})$ and noting that $\mathrm{P}(x_{i,\mathrm{d}} = v_{\tilde{r}}) = \pi_{\tilde{r}}$, by assumption (ii), we get $c = \{\sum_{\tilde{r} \in R(v_{r,i,\mathrm{o}})} \pi_{\tilde{r}}\}^{-1}$. It follows that

$$\mathrm{P}(x_{i,\mathrm{d,m}} = v_{\tilde{r},i,\mathrm{m}}|x_{i,\mathrm{d,o}} = v_{r,i,\mathrm{o}}) = \frac{\pi_{\tilde{r}}}{\sum_{r' \in R(v_{r,i,\mathrm{o}})} \pi_{r'}}, \quad (8.22)$$

$\tilde{r} \in R(v_{r,i,\mathrm{o}})$. The conditional density $f(x_{i,\mathrm{d,m}}|x_{i,\mathrm{d,o}})$ is given by the right side of (8.22) with $v_{r,i,\mathrm{o}}$ replaced by $x_{i,\mathrm{d,o}}$ and $\tilde{r}$ being the $\tilde{r} \in R(x_{i,\mathrm{d,o}})$ such that $v_{\tilde{r},i,\mathrm{m}} = x_{i,\mathrm{d,m}}$. The sample $x_{i,\mathrm{d,m}}$ is drawn from the conditional distribution such that it has the probability given by the right side of (8.22), with $v_{r,i,\mathrm{o}}$ replaced by $x_{i,\mathrm{d,o}}$, of taking the value $v_{\tilde{r},i,\mathrm{m}}, \tilde{r} \in R(x_{i,\mathrm{d,o}})$.

(II) Next, note that, by assumption (iii), we have $x_{i,\mathrm{c}}|x_{i,\mathrm{d}} \sim N(\mu_r, \Omega)$, where r is such that $x_{i,\mathrm{d}} = v_r$. Write $\Omega = (\omega_{kl})_{1 \leq k,l \leq p}$. Then, we have

$$\begin{pmatrix} x_{i,\mathrm{c,m}} \\ x_{i,\mathrm{c,o}} \end{pmatrix} \bigg| x_{i,\mathrm{d}} \sim N \left[\begin{pmatrix} \mu_{r,i,\mathrm{m}} \\ \mu_{r,i,\mathrm{o}} \end{pmatrix}, \begin{pmatrix} \Omega_{i,\mathrm{mm}} & \Omega_{i,\mathrm{mo}} \\ \Omega_{i,\mathrm{om}} & \Omega_{i,\mathrm{oo}} \end{pmatrix} \right],$$

where $\Omega_{i,\mathrm{mm}} = (\omega_{kl})_{k,l \in s_{i,\mathrm{c,m}}}$, $\Omega_{i,\mathrm{mo}} = (\omega_{kl})_{k \in s_{i,\mathrm{c,m}}, l \in s_{i,\mathrm{c,o}}}$, $\Omega_{i,\mathrm{om}} = \Omega'_{i,\mathrm{mo}}$, and $\Omega_{i,\mathrm{oo}} = (\omega_{kl})_{k,l \in s_{i,\mathrm{c,o}}}$ with $s_{i,\mathrm{c,o}} = \{1 \leq k \leq p : x_{i,\mathrm{c},k}$ observed$\}$ ($x_{i,\mathrm{c},k}$ is the kth component of $x_{i,\mathrm{c}}$) and $s_{i,\mathrm{c,m}} = \{1, \dots, p\} \setminus s_{i,\mathrm{c,o}}$. It follows (e.g., Jiang (2007), Appendix C.1), that

$$x_{i,\mathrm{c,m}}|x_{i,\mathrm{c,o}}, x_{i,\mathrm{d}} \sim N\{\mu_{r,i,\mathrm{m}} + \Omega_{i,\mathrm{mo}}\Omega_{i,\mathrm{oo}}^{-1}(x_{i,\mathrm{c,o}} - \mu_{r,i,\mathrm{o}}),$$

$$\Omega_{i,\mathrm{mm}} - \Omega_{i,\mathrm{mo}}\Omega_{i,\mathrm{oo}}^{-1}\Omega_{i,\mathrm{om}}\}. \quad (8.23)$$

Denote the mean vector and covariance matrix of the multivariate normal distribution on the right side of (8.23) by $\mu_{r,i,\mathrm{c,m}}$ and $\Omega_{i,\mathrm{c,m}}$, respectively. Then, we have

$$f(x_{i,\mathrm{c,m}}|x_{i,\mathrm{c,o}}, x_{i,\mathrm{d}})$$

$$\propto \exp\left\{-\frac{1}{2}(x_{i,\mathrm{c,m}} - \mu_{r,i,\mathrm{c,m}})'\Omega_{i,\mathrm{c,m}}^{-1}(x_{i,\mathrm{c,m}} - \mu_{r,i,\mathrm{c,m}})\right\}, \quad (8.24)$$

where r is such that $x_{i,d} = v_r$.

(III) Finally, by (iv), we have $y_{i,m}|x_i \sim N(x'_{i,M}\beta_M, \sigma^2)$, hence

$$f_M(y_{i,m}|x_i) \propto \exp\left\{ -\frac{(y_{i,m} - x'_{i,M}\beta_M)^2}{2\sigma^2} \right\}. \tag{8.25}$$

In conclusion, we can choose (after dropping a constant term) $g(z) =$

$$\left\{ \frac{\pi_{\tilde{r}}}{\sum_{r' \in R(x_{i,d,o})} \pi_{r'}} \right\} \exp\left\{ -\frac{1}{2}(x_{i,c,m} - \mu_{r,i,c,m})' \Omega^{-1}_{i,c,m}(x_{i,c,m} - \mu_{r,i,c,m}) \right.$$

$$\left. -\frac{(y_{i,m} - x'_{i,M}\beta_M)^2}{2\sigma^2} \right\}, \tag{8.26}$$

where $\tilde{r}$ is such that $\tilde{r} \in R(x_{i,d,o})$ and $v_{\tilde{r},i,m} = x_{i,d,m}$, and r is such that $x_{i,d} = v_r$. The sampling from g consists of three steps: (I) draw $x_{i,d,m}$ from the distribution that has the probability equal to $\pi_{\tilde{r}}\{\sum_{r' \in R(x_{i,d,o})} \pi_{r'}\}^{-1}$ of taking the value $v_{\tilde{r},i,m}$ for $\tilde{r} \in R(x_{i,d,o})$; (II) given the $x_{i,d,m}$ drawn, draw $x_{i,c,m}$ from the multivariate normal distribution in (8.23), where r is such that $x_{i,d} = v_r$; (III) given the $x_{i,d,m}, x_{i,c,m}$ drawn, draw $y_{i,m}$ from $N(x'_{i,M}\beta_M, \sigma^2)$.

If any of the components $x_{i,d,m}$, $x_{i,c,m}$, or $y_{i,m}$ are empty, we simply skip the corresponding step(s) (I, II, or III).

The initial estimates of $\mu_r, 1 \le r \le s, \Omega, \pi_r, 1 \le r \le s$ are $\hat{\mu}_r^{(0)} = n_{r,o}^{-1}\sum_{i \in I_{r,o}} x_{i,c}$, $1 \le r \le s$, where $I_{r,o} = \{1 \le i \le n : x_i$ is observed and $x_{i,d} = v_r\}$, and $n_{r,o} = |I_{r,o}|$;

$$\hat{\Omega}^{(0)} = \frac{1}{n_o} \sum_{r=1}^{s} \sum_{i \in I_{r,o}} \{x_{i,c} - \hat{\mu}_r^{(0)}\}\{x_{i,c} - \hat{\mu}_r^{(0)}\}',$$

where $I_o = \cup_{r=1}^{s} I_{r,o}$ and $n_o = |I_o|$, and $\hat{\pi}_r^{(0)} = \#\{1 \le i \le n : x_{i,d}$ observed and $x_{i,d} = v_r\}/\#\{1 \le i \le n : x_{i,d}$ observed$\}, 1 \le r \le s$. Furthermore, the initial estimate of β_f is the LS estimate based on the all-observed data, that is,

$$\hat{\beta}_f^{(0)} = (X'_{ao}X_{ao})^{-1}X'_{ao}y_{ao},$$

where $X_{ao} = (x'_{f,i})_{i \in I_{ao}}$ with $I_{ao} = \{1 \le i \le n : x_{f,i}, y_i \text{ observed}\}$, and $y_{ao} = (y_i)_{i \in I_{ao}}$. The initial estimate of σ^2 is

$$(\hat{\sigma}^2)^{(0)} = \frac{|y_{ao} - X_{ao}\hat{\beta}_f^{(0)}|^2}{|I_{ao}| - p - q}.$$

Next we consider E-MS in conjunction with the AF (Section 3). As noted, the basic element is the EMAF, and the E-MS is implemented via the following algorithm:

a. Take $M_0 = M_f$ as the initial model, and the same initial estimates as given above.

b. For any candidate model M, let $Q(M) = S_0 - S_1' S_2^{-1} S_1$, where S_0 is the same as that below (18), and $S_j, j = 1, 2$ are the same as those below (18) with $x_{i,f}$ replaced by $x_{i,M}$. Note that the conditional expectation, E_f, will be done by the conditional sampling method, described above, with $M = M_0$. Run the AF procedure as described in the second paragraph of Section 2, with $Q(M)$ being the measure of lack-of-fit, to select the optimal model. Denote the optimal model by $\hat{M}$. Let $\hat{\beta} = S_2^{-1} S_1$, where $S_j, j = 1, 2$ are given below (8.19) with $x_{i,f}$ replaced by $x_{i,\hat{M}}$.

c. Let $\hat{\sigma}^2$ be given by (8.19), where $S_j, j = 0, 1, 2$ are given below (8.19) with $x_{i,f}$ replaced by $x_{i,\hat{M}}$. Also, let $\hat{\mu}_r, 1 \leq r \leq s, \hat{\Omega}, \hat{\pi}_r, 1 \leq r \leq s$ be given by (8.20) (note that these depend only on $M_0 = M_f$, but not on $\hat{M}$).

d. Replace M_0 by $\hat{M}$, and the initial estimates by $\hat{\beta}, \hat{\sigma}^2, \hat{\mu}_r, 1 \leq r \leq s, \hat{\Omega}, \hat{\pi}_r, 1 \leq r \leq s$. Repeat steps b and c. Note that, now, the E_f is replaced by $E_{\hat{M}}$, evaluated by the conditional sampling method, as described above, with $M = \hat{M}$.

e. Update the model and parameters iteratively until convergence.

The convergence of the E-MS algorithm, as mentioned in the last sentence, is a key theoretical issue that we shall address in Chapter 9.

8.5 Two simulated examples

In this section, we present two simulation studies on empirical performance of EMAF and E-MS. The first study considers EMAF in the backcross experiment (Example 8.2); the second study considers comparison of EMAF, E-MS, and other model selection strategies in the case of linear regression (Example 8.3).

8.5.1 *EMAF in backcross experiment*

Following Example 8.2, we consider a special case with $q = 6$ and $r = 5$, so there are 5 chromosomes with 6 markers on each chromosome. The sample size is either $n = 250$ or $n = 500$. There are 6 true QTLs, which are located at markers 1, 2, 3 on chromosome 1, markers 1, 2 on chromosome 2, and marker 1 on chromosome 3. The coefficients at the true markers are all equal to 1. The true values for σ^2 and θ are 1 and 0.1, respectively. The complete data are generated as follows: First generate the Markov chain X_f with $\theta = 0.1$; then generate e from $N(0, I_n)$; let $y = X_{opt}(1, 1, 1, 1, 1, 1)' + e$, where X_{opt} has 6 columns corresponding to markers 1, 2, 3 on chromosome 1, markers 1, 2 on chromosome 2, and marker 1 on chromosome 3. Next, we generate a 10% random sample, I_m, from $\{1, \ldots, n\}$, and a 10% random sample, M_{jk}, from $\{1, \ldots, n\}$, $1 \le k \le r, 1 \le j \le q$. So, there are $qr + 1$ 10% random samples, corresponding to the indexes of the missing data. This leaves less then 4% of the complete-data records, on average (similar to the backcross data obtained by the researchers at UC-Riverside; see Example 8.2). Let $I_o = \{1, \ldots, n\} \setminus I_m$ and $O_{jk} = \{1, \ldots, n\} \setminus M_{jk}, 1 \le k \le r, 1 \le j \le q$. The subsets I's, M's and O's are fixed throughout the simulations. The observed data are $y_i, i \in I_o$, and $x_{ijk}, i \in O_{jk}, 1 \le k \le r, 1 \le j \le q$.

We apply EMAF using RF, as described in Section 4, in which a strategy called wild bootstrapping was introduced. It was shown that the RF is relatively robust with respect to different parametric bootstrap methods. To see if a similar behavior holds in the presence of missing data, we carry out a simulation study, under different parametric bootstrap schemes. More specifically, the double-exponential distribution, or Laplace distribution, has been motivated for both historical and practical reasons. On the one hand, it has historically been considered as a firm candidate (though not unique) to substitute the usual normality assumption; on the other hand, it is the most preferred under certain special circumstances (e.g., with environment radiological measures due to their exponential decay). It also highlights the median-based inferential methods which are considered more robust than the mean-based methods. Due to these considerations, we have considered bootstrapping under the double-exponential distribution, instead of under the normal distribution. Thus, when the errors are actually normal, the double-exponential strategy may be consider as wild bootstrapping (Subsection 4.2.2).

Under the sample sizes $n = 250$ and $n = 500$, results based on 100 simulation runs are presented in Table 8.2. Reported are empirical true-positive (TP), that is, proportion of times out of the total number of simulation runs that the method identifies all of the true QTLs, and nothing else; empirical mean number of correctly identified QTLs (MC); and empirical mean number of incorrectly identified QTLs (MIC), with the corresponding empirical s.d.'s in the parentheses. Here, the bootstrap scheme is either under normal error (Normal), or double-exponential error (Laplace). The true

Table 8.2 **EMAF under Different Bootstrap Schemes.**

n	Bootstrap Scheme	TP	MC (s.d.)	MIC (s.d.)
250	Normal	0.85	5.86 (0.37)	0.02 (0.14)
250	Laplace	0.86	5.89 (0.37)	0.10 (0.30)
500	Normal	0.98	5.99 (0.10)	0.02(0.14)
500	Laplace	0.99	5.99 (0.10)	0.01(0.10)

error is normal; thus Laplace bootstrapping is considered "wild" in a way similar to Subsection 4.2.2. As can be seen, there is no essential difference between the two bootstrap schemes; in particular, the wild bootstrapping works just as well as bootstrapping under the true error distribution.

8.5.2 *Linear regression: Comparison of strategies*

Following Example 8.3, suppose that the candidate covariates consist of two continuous variables and two indicator variables. So, x_1, x_2 are continuous (with $p = 2$) and x_3, x_4 are indictors (0 or 1). Thus, $x_i = (x_{i1}, x_{i2}, x_{i3}, x_{i4})'$ with $x_{i,c} = (x_{i1}, x_{i2})'$ and $x_{i,d} = (x_{i3}, x_{i4})'$. The distinct possible values for $x_{i,d}$ are $v_1 = (0, 0)'$, $v_2 = (0, 1)'$, $v_3 = (1, 0)'$, and $v_4 = (1, 1)'$. Assumption (iii) of Example 8.3 means that there are 2×1 vectors $\mu_1, \mu_2, \mu_3, \mu_4$ and 2×2 covariance matrix Ω such that, given $x_{i,d} = v_r$, $x_{i,c} \sim N(\mu_r, \Omega)$, $r = 1, 2, 3, 4$.

Simulations are run with the sample size $n = 100$ and the true model being x_1 and x_3. The true parameters are $\beta_1 = \beta_3 = \sigma^2 = 1$. The true μ_r is the same as v_r, $r = 1, 2, 3, 4$; and the true Ω is the 2×2 identity matrix. After the complete data is generated, we randomly select a subset of indexes from $\{1, \ldots, n\}$ for the response as well as for each of the candidate predictors, which correspond to the missing data, so that $100 p_m \%$ of the data are missing for the response and each of the candidate predictors. Here we consider two cases: $p_m = 0.1$ and $p_m = 0.2$.

We study performance of E-MS with IF, as described in Section 5. It is known that the latter may suffer from the dominant factor effect (see Section 5.3). Namely, although the true coefficients for the true predictors (x_1 and x_3) are both equal to 1, it turns out that the continuous predictor is a dominant factor. Thus, with a moderate sample size, such as in the current case, the IF tends to overwhelmingly select x_1 at dimension 1, leading to a potentially underfitting model. To overcome such a problem, we consider the following modification of the IF to make it more "aggressive". Let α be a chosen number between 0 and 1. Let d^* be the selected dimension by IF and p^* be the corresponding maximum empirical probability (that a given model is selected at the dimension). Let p_1^* be the largest maximum empirical probability corresponding to a dimension greater than d^* (thus, $p_1^* < p^*$). Let d_1^* be the corresponding dimension to p_1^*. If $p_1^* \geq (1 - \alpha)p^*$, then d_1^* is selected instead of d^*. It is clear that the modified IF is more in favor of a "larger" model, and in this sense it is more aggressive.

We run a same-data comparison of E-MS with a number of different procedures. The first is an IF version of EMAF. Namely, we first run the E-M algorithm to obtain the parameter estimates under the full model. We then generate (parametric) bootstrap samples under the full model. As noted (see Section 8.3), the best part of this procedure is that, when one generates the bootstrap samples, one generates complete data rather than data with missing values. We then apply the modified IF, as described above, to the bootstrapped data. We call such a procedure EMIF. In our simulation study, we consider three different values of α: $\alpha = 0$ (corresponding to IF without the modification), $\alpha = 0.1$, and $\alpha = 0.5$, for EMIF as well as each of the comparing procedures described below, except the E-MS. Of course, this raises a question on what α is the best, which one may not know in practice. On the other hand, the E-MS seems to have some advantage in this regard. The idea is to start with a relatively large α (say, $\alpha = 0.5$), in order to be more conservative in dropping the predictors, and gradually reduces α as the iteration progresses. More specifically, we begin with $\alpha = 0.5$; with each iteration, we reduce α by half, until convergence.

In addition to EMIF and E-MS, two other methods are also included in our comparison. One is IF based on the complete-record-only analysis (CRNIF); the other is IF with the missing data replaced by the imputed data (IMIF). The latter is based on a method of multivariate imputation developed by van Buuren *et al.* (2005), implemented in the R package, aregImpute(). As a comparison, we have also considered IF based on the complete data that were generated before the missing values were taken out (CDIF). The latter is, of course, not possible in a practical situation, but it would be interesting to see how much loss of efficiency there is for a method, if any, compared to the "gold standard". Note that all of the comparing methods are in conjunction with the IF; the only difference is how the missing data are handled. The results based on 100 simulation runs are presented in Tables 8.3 and 8.4, where OF standards for overfitting, that is, the empirical probability that the selected model includes all of the true predictors, plus at least one extraneous predictors; UF standards for underfitting, that is, the empirical probability that the selected model misses at least one true predictors [but may include extraneous predictor(s)

Table 8.3 **Summary of Performance** ($p_m = 0.1$).

Method	α	TP	OF	UF	MC	s.d.	MIC	s.d.
CRNIF	0	0.67	0.00	0.33	1.67	0.47	0.00	0.00
	0.1	0.85	0.00	0.15	1.85	0.36	0.00	0.00
	0.5	0.73	0.27	0.00	2.00	0.00	0.27	0.45
IMIF	0	0.61	0.00	0.39	1.61	0.49	0.00	0.00
	0.1	0.88	0.00	0.12	1.88	0.33	0.00	0.00
	0.5	0.72	0.28	0.00	2.00	0.00	0.28	0.45
EMIF	0	0.66	0.00	0.34	1.66	0.48	0.00	0.00
	0.1	0.88	0.00	0.12	1.88	0.33	0.00	0.00
	0.5	0.75	0.25	0.00	2.00	0.00	0.25	0.44
CDIF	0	0.71	0.00	0.29	1.71	0.46	0.00	0.00
	0.1	0.92	0.00	0.08	1.92	0.27	0.00	0.00
	0.5	0.69	0.31	0.00	2.00	0.00	0.31	0.46
E-MS		0.98	0.01	0.01	1.99	0.10	0.01	0.10

Table 8.4 **Summary of Performance** ($p_m = 0.2$).

Method	α	TP	OF	UF	MC	s.d.	MIC	s.d.
CRNIF	0	0.56	0.00	0.44	1.56	0.50	0.01	0.10
	0.1	0.73	0.07	0.20	1.80	0.40	0.09	0.29
	0.5	0.68	0.31	0.01	1.99	0.10	0.32	0.47
IMIF	0	0.59	0.00	0.41	1.59	0.49	0.01	0.10
	0.1	0.85	0.00	0.15	1.85	0.36	0.01	0.10
	0.5	0.81	0.19	0.00	2.00	0.00	0.19	0.39
EMIF	0	0.65	0.00	0.35	1.65	0.48	0.00	0.00
	0.1	0.88	0.00	0.12	1.88	0.33	0.00	0.00
	0.5	0.82	0.17	0.01	1.99	0.10	0.18	0.39
CDIF	0	0.75	0.00	0.25	1.75	0.44	0.00	0.00
	0.1	0.94	0.00	0.06	1.94	0.24	0.00	0.00
	0.5	0.74	0.26	0.00	2.00	0.00	0.26	0.44
E-MS		0.95	0.01	0.04	1.96	0.20	0.03	0.17

as well]; other performance measures are the same as before (corresponding s.d.'s to the right, for MC and MIC).

It is clear in this comparison that, overall, E-MS outperforms not only all of the methods that are practically feasible (CRNIF, IMIF, EMIF), but also the "gold standard" (CDIF) that is practically infeasible. In fact, in terms of the overall performance, the order seems to be (from best to worst) E-MS, CDIF, EMIF, IMIF, CRNIF. Of course, it is not surprising that CRNIF takes the last place, but what seems a little unexpected is that E-MS even (slightly) outperforms CDIF. An explanation for this is that the performance of CDIF still suffers, to some extend, the dominant factor effect (see Section 5.3), but E-MS is able to overcome this (see below). A key to this "super-performance" is α, which may be viewed as a tuning parameter.

It appears that the best α for CRNIF, IMIF, EMIF, and CDIF is somewhere between 0.1 and 0.5. Of course, in the simulation study we could explore this best value, but it would not be possible in practice. On the other hand, the E-MS seems to be able to get the best out of the "α-business" during its iterations. By the way, in all of the simulation runs, the E-MS converged in 2-3 iterations. Also, it seems that the performance of IMIF, EMIF, and E-MS are not affected much by the increase of p_m. This is a bit surprising, as larger p_m means less observed data. In fact, with $p_m = 0.1$, one expects about 59% complete data records; with $p_m = 0.2$, the % of the complete data records drops to less than 33%. On the other hand, it takes about twice the computing time to run the E-MS for $p_m = 0.2$, compared to $p_m = 0.1$. This is reasonable, as more data are missing under $p_m = 0.2$; therefore, the conditional expectations, which have to be dealt with via the Monte-Carle method (see Example 8.3), need to be evaluated more often than under $p_m = 0.1$.

8.6 Missing data mechanisms

In a way, there are three cases that the missing data mechanisms (MDM) may be involved. The first case, case I, is that the MDM is known, which is rarely the case in practice; the second case, case II, is that the MDM is also of interest, and subject to model selection; the third case, case III, is that the MDM is unknown, but is not of interest; in other words, in case III, there is an underlying MDM, but the latter is something that one wishes to avoid dealing with. In our experience, the third case is encountered most frequently in practice.

The presented E-MS method applies to cases I and II without any change. This is because, in those cases, the observed data include both y_{obs}, which is what we normally call "the data" without considering the MDM, and the missing data indicators, m_{ind}. In other words, the full (observed) data is (y_{obs}, m_{ind}). Under either case I or case II, one has a complete specification of the distribution of (Y_{obs}, M_{ind}), that is,

$$f(y_{obs}, m_{ind}|\theta, \psi) = \int f(y|\theta)f(m_{ind}|y, \psi)dy_{mis}. \qquad (8.27)$$

The first factor inside the integral on the right side of (8.27) corresponds to the distribution of the complete data, $Y = (Y_{obs}, Y_{mis})$, where Y_{mis} represents the missing data; the second factor, $f(m_{ind}|y, \psi)$, corresponds to the MDM. Here θ and ψ denote the parameter vectors that are involved in the distribution of Y and the MDM, respectively. Therefore, from a methodology point of view, there is nothing new and (8.27) is just a special case to which the E-MS applies, that is, a set of data and a distribution for the data under an assumed model, a part of which is the MDM.

A more challenging case seems to be case III, in which one is interested in the model on Y only, and would avoid dealing with the MDM if possible. As noted, this case is encountered most frequently in practice. Of course, one may always consider some candidate models for the MDM, and treat the case the same way as case II; once a joint model is selected, one simply takes the part regarding the distribution of Y, which is of main interest. The question is: How does the latter approach compare to the E-MS that focuses on the Y model only? Another related question is: How is the performance of the E-MS, which ignores the MDM, affected by the true underlying MDM? In this section, we address these questions from a theoretical standpoint, and refer relevant empirical results, for the case of E-MS with BIC, to Jiang *et al.* (2014).

We refer to Rubin (1976) and Little and Rubin (2002) for the well known theory about MDM, including the notions of MCAR, MAR, NMAR; and ignorable and non-ignorable MDM. According to Little and Rubin (2002) (sec. 6.2), the frequentist's methods of inference that ignore the MDM are still valid, even if the MDM is non-ignorable, although there may be a loss of efficiency. It follows that the E-MS, as a frequentist's method, is valid even without considering the MDM; on the other hand, there may be a loss of efficiency in terms of model selection performance. Furthermore, if the true MDM is ignorable, there is no loss of efficiency in any likelihood-based inference, including model selection, by ignoring the MDM. Therefore, the case of interest is when the MDM is non-ignorable.

For simplicity, let us assume that the observations Y_i are independent Gaussian with mean $\mathrm{E}_{M,\theta_M}(Y_i)$, where M indicates the assumed model for the mean, and θ_M the vector of parameters under M, and unknown variance σ^2. Consider selection of M using the E-MS that is based on the measure of lack-of-fit, $\mathrm{E}_c(Q_M|y_{\mathrm{obs}})$, where y_{obs} denotes the observed data, and E_c denotes conditional expectation under the current model, M_c, and parameters under M_c, and

$$Q_M = \sum_{i=1}^{n} \{Y_i - \mathrm{E}_{M,\theta_M}(Y_i)\}^2. \tag{8.28}$$

Furthermore, suppose that the criterion of selecting the model within the fence does not depend on the MDM. For example, if the minimum-dimension criterion is used to select the model within the fence, then the criterion is determined by $|M|$, the dimension of the model, which does not depend on the MDM. The derivation below requires, of course, some regularity conditions [e.g., Jiang *et al.* (2002)]; however, we shall bypass these technical conditions and focus on the insight of the result.

Let $m_{\text{ind},i}$ denote the missing data indicator. Then, we have

$$E_c(Q_M|y_{\text{obs}}) = \sum_{i=1}^{n} \{y_i - E_{M,\theta_M}(Y_i)\}^2 1_{(m_{\text{ind},i}=0)}$$

$$+ \sum_{i=1}^{n} E_c\{Y_i - E_{M,\theta_M}(Y_i)\}^2 1_{(m_{\text{ind},i}=1)}. \qquad (8.29)$$

Suppose that the current model is correct, but not necessarily optimal. For example, if the space of candidate models includes a true model, then the full model, M_f, is correct, but not necessarily optimal in that it may include extraneous variables. Furthermore, suppose that the current estimator of parameters is consistent. Then, the conditional expectation, E_c, can be replaced by the true conditional expectation, E, resulting a difference that is of lower order. Another situation is when the E-MS results in consistent model selection (see Chapter 9). Then, asymptotically, one can replace E_c by E. Furthermore, by Theorem 2 of Jiang *et al.* (2011a), the minimizer of (8.29), with E_c replaced by E, $\hat{\theta}_M$, converges in probability to some limiting vector, say, θ_M, and this is true regardless whether M is a correct model. Thus, by considering the leading term, we can focus on (8.29) with E_c replaced by E, and θ_M being the limiting vector. Let $P(M_{\text{ind},i} = 1|y) = 1 - h(y_i)$ be the true underlying MDM; in other words, $h(y_i) = P(M_{\text{ind},i} = 0|y)$, where y is the complete data. Define $c_i = E\{Y_i h(Y_i)\}/E\{h(Y_i)\}$ (again, E without subscript represents the true expectation). Note that

$$\{Y_i - E_{M,\theta_M}(Y_i)\}^2 1_{(M_{\text{ind},i}=0)} = (Y_i - c_i)^2 1_{(M_{\text{ind},i}=0)}$$

$$+ 2(Y_i - c_i)\{c_i - E_{M,\theta_M}(Y_i)\} 1_{(M_{\text{ind},i}=0)}$$

$$+ \{c_i - E_{M,\theta_M}(Y_i)\}^2$$

$$= \xi_i + 2\eta_i + \zeta_i. \qquad (8.30)$$

It can be shown that $E(\eta_i) = 0$ (Exercise 8.8). Thus, by the law of large numbers (LLN), we have $\sum_{i=1}^{n} \eta_i = o_P(n)$ [e.g., Jiang (2010), ch. 3]. It follows that

$$\sum_{i=1}^{n} \{Y_i - E_{M,\theta_M}(Y_i)\}^2 1_{(M_{\text{ind},i}=0)}$$

$$= \sum_{i=1}^{n} (Y_i - c_i)^2 1_{(M_{\text{ind},i}=0)} + \sum_{i=1}^{n} \{c_i - E_{M,\theta_M}(Y_i)\}^2 1_{(M_{\text{ind},i}=0)} + \delta$$

$$= \sum_{i=1}^{n} \{c_i - E_{M,\theta_M}(Y_i)\}^2 1_{(M_{\text{ind},i}=0)} + \delta_1,$$

where δ is a lower-order term, and δ_1 is a sum of a term that is not model-dependent and a lower-order term. Similarly, it can be shown that

$$\sum_{i=1}^{n} E_c \{Y_i - E_{M,\theta_M}(Y_i)\}^2 1_{(M_{\text{ind},i}=1)}$$

$$= \sum_{i=1}^{n} \{E(Y_i) - E_{M,\theta_M}(Y_i)\}^2 1_{(M_{\text{ind},i}=1)} + \delta_2,$$

where and δ_2 is a sum of a term that is not model-dependent and a lower-order term. Thus, by (8.29), we have

$$E_c(Q_M|y_{\text{obs}}) = \sum_{i=1}^{n} \{c_i - E_{M,\theta_M}(Y_i)\}^2 1_{(M_{\text{ind},i}=0)}$$

$$+ \sum_{i=1}^{n} \{E(Y_i) - E_{M,\theta_M}(Y_i)\}^2 1_{(M_{\text{ind},i}=1)} + \delta,$$

where δ is a sum of terms that are not model-dependent and lower-order terms. Thus, by taking the expectation, we obtain

$$E\{E_c(Q_M|Y_{\text{obs}})\} = \sum_{i=1}^{n} \{c_i - E_{M,\theta_M}(Y_i)\}^2 E\{h(Y_i)\}$$

$$+ \sum_{i=1}^{n} \{E(Y_i) - E_{M,\theta_M}(Y_i)\}^2 [1 - E\{h(Y_i)\}]$$

$$+ \delta, \tag{8.31}$$

where δ consists of lower-order terms, or terms that do not depend on M. Let M_{opt} denote the optimal model. Then, because (8.31) holds for every M, by letting $M = M_{\text{opt}}$, we have

$$E\{E_c(Q_{M_{\text{opt}}}|Y_{\text{obs}})\} = \sum_{i=1}^{n} \{c_i - E(Y_i)\}^2 E\{h(Y_i)\} + \delta_{\text{opt}}, \tag{8.32}$$

where δ_{opt} consists of terms that are not model-dependent and lower-order terms. Note that, when $M = M_{opt}$, E_{M,θ_M} becomes E, and the second term on the right side of (8.31) disappears. By taking the difference between (8.31) and (8.32), we get

$$E\{E_c(Q_M|Y_{obs})\} - E\{E_c(Q_{M_{opt}}|Y_{obs})\}$$
$$= \sum_{i=1}^{n}\{c_i - E_{M,\theta_M}(Y_i)\}^2 E\{h(Y_i)\} + \sum_{i=1}^{n}\{E(Y_i) - E_{M,\theta_M}(Y_i)\}^2 E\{g(Y_i)\}$$
$$- \sum_{i=1}^{n}\{c_i - E(Y_i)\}^2 E\{h(Y_i)\} + \delta,$$

where δ has the same meaning as in (8.31). Furthermore, note that

$$\sum_{i=1}^{n}\{E(Y_i) - E_{M,\theta_M}(Y_i)\}^2 E\{g(Y_i)\} = c - \sum_{i=1}^{n}\{E(Y_i) - E_{M,\theta_M}(Y_i)\}^2 E\{h(Y_i)\},$$

where c does not depend on the MDM. Also note that

$$\{c_i - E_{M,\theta_M}(Y_i)\}^2 - \{E(Y_i) - E_{M,\theta_M}(Y_i)\}^2 - \{c_i - E(Y_i)\}^2$$
$$= 2\{c_i - E(Y_i)\}\{E(Y_i) - E_{M,\theta_M}(Y_i)\},$$
$$\{c_i - E(Y_i)\}\{E(Y_i) - E_{M,\theta_M}(Y_i)\}E\{h(Y_i)\}$$
$$= [E\{Y_i h(Y_i)\} - E(Y_i)E\{h(Y_i)\}]\{E(Y_i) - E_{M,\theta_M}(Y_i)\}$$
$$= cov\{Y_i, h(Y_i)\}\{E(Y_i) - E_{M,\theta_M}(Y_i)\}$$

[recall the definition of c_i above (8.30)]. Also, the previous derivation, and the LLN, show that $E\{E_c(Q_M|Y_{obs})\} - E\{E_c(Q_{M_{opt}}|Y_{obs})\}$ is the leading term for the difference in (8.29) between M and M_{opt}. It follows that

$$\text{difference in (8.29) between } M \text{ and } M_{opt}$$
$$= E\{E_c(Q_M|Y_{obs})\} - E\{E_c(Q_{M_{opt}}|Y_{obs})\} + \delta_1$$
$$= 2\sum_{i=1}^{n} cov\{Y_i, h(Y_i)\}\{E(Y_i) - E_{M,\theta_M}(Y_i)\} + \delta_2, \qquad (8.33)$$

where δ_1 denotes terms of lower-order, and δ_2 consists of terms of lower-order, or terms that do not depend on the MDM. (8.33) is a key result that shows how the performance of the E-MS is influenced by the MDM through its leading term, namely, the larger this term (i.e., more positive), the easier to distinguish a non-optimal model from the optimal one. It is interesting to note that the leading term is a sum of products, where the first factor of the product, $cov\{Y_i, h(Y_i)\}$, depends on the MDM but not on

M, while the second factor of the product, $E(Y_i) - E_{M,\theta_M}(Y_i)$, depends on M but not on the MDM.

Expression (8.33) may help to explain, for example, some apparent "super-performances" of E-MS observed in Jiang *et al.* (2014), namely, that under certain MDMs, E-MS with BIC can out-perform BIC based the complete data. The latter include both the observed data and the unobserved missing data, had the latter been observed. The results are not presented here because those are regarding E-MS with BIC, rather than the fence. Nevertheless, from (8.33), it is seen that if the MDM "works in the right direction", the difference in the measure of lack-of-fit between an incorrect candidate model and the optimal model can be "amplified"; therefore, it is easier to separate an incorrect model from the optimal one. To see a more intuitive explanation, consider the following hypothetical example. Suppose that the data are measurements of a certain lung function, and that one intends to build a model that associates the lung function to status of smoking. A patient of lung cancer who was a smoker has died; therefore, measurements of the lung function are missing for the patient. But, guess what? Indication that the data are missing due to the death might be an even stronger evidence for building the model than having the measurements, because the subject has died (due to the cancer).

8.7   Real data example

We use a real-data example to further illustrate the E-MS method. Recall the data set obtained by the UC-Riverside researchers mentioned earlier (e.g., Section 8.1; Example 8.2). The gene expression data were originally published by Luo *et al.* (2007). The phenotypic values of eight quantitative traits of barley were published by Hayes *et al.* (1993). Detailed description of the experiment can be found in the latter reference, which involved 150 double haploid (DH) lines derived from the cross of two spring barley varieties, Morex and Steptoe. The DH lines are considered as the subjects here. In all there were 495 SNP markers on seven chromosomes that are under investigation. As mentioned, there are significant missing values in the data so that only 4 of the 150 subjects have complete genotype records. On the other hand, there are no missing values in the phenotypic data.

We consider a Markov-chain model as in Example 8.2. However, the high-dimensional nature of the data presents a problem for the direct application of the E-MS, because the total number of markers (495) is much larger than the sample size ($n = 150$). More specifically, the least squares (LS) fit is unfeasible when the number of predictors is larger than the sample size. To overcome this difficulty, we use the following idea of *conditional modeling*, described under a more general setting.

Suppose that, conditional on $X = (x_i')_{1 \leq i \leq n}$, one has a linear regression $Y = X\beta + \epsilon$, where $Y = (Y_i)_{1 \leq i \leq n}$ are the observations, and $\epsilon = (\epsilon_i)_{1 \leq i \leq n}$ are the errors such that the components of ϵ are independent with mean 0, and ϵ is independent of X. Furthermore, suppose that $X = [X_{(1)} \ X_{(2)}]$ with $X_{(r)} = (X_{ir}')_{1 \leq i \leq n}, r = 1, 2$ such that $X_{(1)}, X_{(2)}$ are independent [e.g., Broman and Speed (2002)]. Then, it is easy to show that $X_{(1)}$ is independent of $[X_{(2)}, \epsilon]$. Note that we can express the regression model as

$$Y = X_{(1)}\beta_1 + X_{(2)}\beta_2 + \epsilon.$$

Without loss of generality, we assume that $X_{(1)}\beta_1$ does not involve an intercept [which, if exist, belongs to $X_{(2)}\beta_2$].

Now suppose that $X_{i2}, i = 1, \ldots, n$ are independent, and that $\mathrm{E}(X_{i2})$ does not depend on i. Then, $\mathrm{E}(X_{i2}'\beta_2 + \epsilon_i) = \mathrm{E}(X_{i2})'\beta_2$ is a constant, say, β_0. Let $e_i = X_{i2}'\beta_2 + \epsilon_i - \beta_0$. It is easy to show that $e_i, i = 1, \ldots, n$ are independent with $\mathrm{E}(e_i) = 0$, and $Y = [1_n \ X_{(1)}](\beta_0 \ \beta_1')' + e$, e being independent of $[1_n \ X_{(1)}]$. In other words, conditional on $X_{(1)}$, we, once again, have a standard linear regression model (i.e., the errors are independent with mean zero, and independent with the predictors).

The point is that $X_{(1)}$ can be of much lower dimension than X. For the barley cross data, we can let $X_{(1)}$ correspond to markers on any particular chromosome. The number of markers on the 7 chromosomes are 60, 78, 81, 60, 93, 56 and 67, respectively, all of which are smaller than the sample size 150. Within each chromosome, we apply the E-MS in conjuction with the IF (see Subsection 8.5.2). The number of bootstrap samples is chosen as $B = 100$.

It is known that, for high-dimensional data the IF may suffer from the dominant factor effect (see Section 5.3). For the most part, this means that the IF frequency, that is, empirical probability of the most frequently selected model, tends to be in favor of a lower dimensional model than the true model, if the "signals" are relatively weak due to the limited sample size. This problem is dealt with naturally by the E-MS. First we apply the IF, under the full model, that is, all of the markers on a given chromosome, to obtain the IF frequencies at different dimensions, say, $p_1^*, p_2^*, \ldots, p_q^*$, where p_j^* is the IF frequency at dimension j, and q is the total number of markers, for the chromosome. If the frequencies show a "peak", that is, there is a

$1 < j < q$ such that $p_j^* > p_{j-1}^*$ and $p_j^* > p_{j+1}^*$, the E-MS shall continue; otherwise, we conclude that there is no more than one QTL on the chromosome. In the latter case, the highest IF frequecy must take place at the boundary, that is, either at dimension one or at the highest dimension corresponding to all the markers on the chromosome. However, it is unlikely that all the markers are QTLs; therefore, dimension one is chosen, and the E-MS stops.

If the frequency plot show a "peak", and therefore the E-MS is to continue, we first look for the last peak, that is, the highest dimension that corresponds to a peak in order to be conservative. This is similar to the AF (see Section 3.2), where the first significant peak is chosen in order to determine the cut-off for the fence [e.g., Nguyen and Jiang (2012)]. The first peak for the AF corresponds to the last peak for the IF. The markers corresponding to the last peak are selected, the current model is updated, and the updated model is treated as the (new) full model for the next step of iteration. The procedure is repeated until either the updated model is identical to the current model, or no peak is found during the current step; in both cases, the current model is chosen as the final model. For the latter case, when no peak is found, we choose the highest dimension, instead of dimension one as above in the initial step. This is because, at this stage, we have already determined that there are more than one QTLs on the chromosome (the E-MS would not have continued otherwise); furthermore, the highest dimension possibly has been updated, so it no longer corresponds to all of the markers on the chromosome.

The results for the grain protein phenotype are presented in Table 8.5. The results show some consistency with the foundings of Zhan *et al.* (2011). For example, the latter authors found that chromosomes 2, 3, 5 "seem to control more genes than other chromosomes". According to our results, those three chromosomes contain nearly 60% (10 out of 17) of all the QTLs found. In particular, chromosomes 3 and 5 are the top two according to the number of QTLs found. It should be noted that the number of QTLs found on a chromosome is not the only thing that represents the relative importance of the chromosome; the magnitude of the QTL effect is also important. In this application, however, our focus is identification of the QTLs, rather than estimation of the QTL effects.

Table 8.5 **E-MS Results for Grain Protein.**

Chromosome	Marker ID#				Chromosome	Marker ID#			
1	12	13			5	280	285	332	333
2	65	66			6	379	380		
3	184	186	199	200	7	467	470		
4	176								

8.8 Exercises

8.1. Verify (8.1) and (8.2) in Example 8.1.

8.2. Verify expressions (8.3) and (8.4) in Example 8.2.

8.3. In Example 8.2, verify expressions (8.5)–(8.7).

8.4. In Example 8.2, verify the two expressions of MLE, (8.8), and that of MLE based on all of the x's, (8.9).

8.5. Verify the conditional expectations (8.13) and (8.14). Also, show that the solution to the equation that lets the derivatives of (8.10) equal to zero is given by (8.15) and (8.16).

8.6. In Example 8.3, verify the expressions of parameter updates for β, σ^2, μ_r, π_i, and Ω, (8.19) and (8.20).

8.7. In example 8.3, show that one can sample from the joint conditional distribution of the data (including y and x) given y_o, x_o, by sampling independently from the conditional distribution for each subject, i; and the latter is given by (8.21).

8.8. Show that, in (8.30), one has $E(\eta_i) = 0$.

Chapter 9

Theoretical Properties

All theory is gray, my friend. But forever green is the tree of life. – John Wolfgang von Goethe (1808), *Faust, First Part*

9.1 Introduction

There have been studies on theoretical properties of the fence and its variations. An important theoretical property of a model selection strategy, including the fence, is *consistency*. This means that, as the sample size, n, increases, the probability of selecting the optimal model goes to one. The term "optimal model" is used here, which, by the classical definition, refers to a true model with the minimum dimension. Consider, for example, the problem of regression variable selection. Suppose that the underlying data-generating process is

$$y = \beta_0 + \beta_1 x_1 + \beta_2 x_2 + \beta_3 x_3 + \epsilon, \tag{9.1}$$

where y is the response; x_1, x_2, x_3 are variables, or predictors, that are linearly associated with y, for which the regression coefficients, $\beta_1, \beta_2, \beta_3$ are non-zero; and ϵ is the regression error. Now suppose that the space of candidate variables include $x_j, j = 1, \ldots, 15$. Thus, for example, $y = \beta_0 + \beta_1 x_1 + \beta_{10} x_{10}$ is an incorrect model, because it excludes the true variables x_2, x_3. On the other hand, $y = \beta_0 + \beta_1 x_1 + \beta_2 x_2 + \beta_3 x_3 + \beta_{10} x_{10}$ is a true model, because it includes all of the true variables, plus one extraneous variable, x_{10}. It is a true model because, if the coefficient β_{10} equals zero, the model reduces to (9.1); however, it is not the optimal model, because the inclusion of x_{10} in unnecessary. In fact, by fitting a model with extraneous variables, one has to estimate the regression coefficients associated with these variables, which are supposed to be zero. For example, the

LS estimators of these (zero) regression coefficients are never zero. For such a reason, a model that includes all of the true variables plus at least one extraneous variables is called an overfitting model. The only optimal model in this case is (9.1), which includes all of the true variables, and nothing else. By consistency, it means that, as $n \to \infty$, the probability that the selected model is (9.1) goes to one.

A different type of consistency property is called *signal-consistency*. The classical definition of consistency, either for parameter estimation [e.g., Jiang (2010)] or for model selection (as described above), assumes that the sample size goes to infinity. In many cases, this assumption may not be realistic. For example, in genome-wide association studies [GWAS; e.g., Hindorff *et al.* (2009)], the number of genetic variants in the genome for certain traits of interest, which corresponds to the x variables, is often much larger than the number of unrelated individuals involved in the study, which corresponds to the sample size n. Jiang *et al.* (2011b) proposed a scenario, in which the sample size is fixed while the "signal" increases. For example, in the above example of regression variable selection, the signals are the absolute values of $\beta_1, \beta_2, \beta_3$, or, more precisely, the minimum of those absolute values. The point is to show that the probability of selecting the optimal model goes to one as the signal increases. Of course, one may not be able to increase the signals in real-life, but a similar statement can also be made about sample size going to infinity. The point is to see if a procedure works perfectly well in an "ideal situation", which we believe is a basic property, just like consistency in the classical sense.

The last part of the theory presented here is regarding (numerical) convergence of the E-MS algorithm (see Section 8.4) as well as consistency (in model selection) of the limit of the E-MS convergence. Such a theory has been established in Jiang *et al.* (2014) for E-MS with generalized information criterion [GIC; Nishii (1984), Shibata (1984)] and E-MS with the fence, but here we shall focus on the latter.

The theory established so far is by no mean complete. For example, although the technical conditions have intuitive interpretations, the conditions are not "clean" in the sense that some of the so-called "regularity conditions" ought to be proved, rather than assumed. However, it seems difficult to prove these conditions, which likely involves asymptotic analysis of the joint distribution of the observed data and bootstrap samples. It is certainly of interest to overcome such technical difficulties, and obtain the asymptotic results under cleaner conditions.

On the other hand, just because the theory is not (yet) perfect, it does not stop one from using the fence methods. In general, one does not have to wait until there is a perfect theory to start using a method; in fact, it is often the opposite–real-life practice is the driving force for theoretical developments. From this point of view, the quote at the beginning of the chapter said it perfectly.

9.2 Consistency of fence, F-B fence and AF

The measure of lack-of-fit, $Q(M, \theta_M; y)$, will be abbreviated as $Q(M, \theta_M)$, $Q(M)$, or simply Q, depending on the occasion. We assume that the following assumptions $A1$–$A4$ hold for each $M \in \mathcal{M}$, the space of candidate models, where θ_M represents a parameter vector at which $\mathrm{E}\{Q(M)\}$ attains its minimum. $\partial Q / \partial \theta_M$ and $\partial \tilde{Q} / \partial \theta_M$, etc., represent partial derivatives of $Q(M)$, with respect to θ_M, evaluated at θ_M and $\tilde{\theta}_M$, respectively. Also, throughout this chapter, we use the words "correct model" (see the first paragraph of Section 9.1) and "true model" interchangeably.

A1. $Q(M)$ is three-times continuously differentiable with respect to θ_M such that $\mathrm{E}(\partial Q / \partial \theta_M) = 0$.

A2. There is a constant B_M such that $|\tilde{\theta}_M| > B_M$ implies $Q(M, \tilde{\theta}_M) > Q(M, \theta_M)$.

A3. The equation $\partial Q(M) / \partial \theta_M = 0$ has an unique solution.

A4. There is a sequence of positive numbers $a_n \to \infty$ and $0 \leq \gamma < 1$ such that the following hold:

$$\frac{\partial Q}{\partial \theta_M} - \mathrm{E}\left(\frac{\partial Q}{\partial \theta_M}\right) = O_\mathrm{P}(a_n^\gamma),$$

$$\frac{\partial^2 Q}{\partial \theta_M \partial \theta'_M} - \mathrm{E}\left(\frac{\partial^2 Q}{\partial \theta_M \partial \theta'_M}\right) = O_\mathrm{P}(a_n^\gamma),$$

$$\liminf \frac{1}{a_n} \lambda_{\min}\left\{\mathrm{E}\left(\frac{\partial^2 Q}{\partial \theta_M \partial \theta'_M}\right)\right\} > 0,$$

$$\limsup \frac{1}{a_n} \lambda_{\max}\left\{\mathrm{E}\left(\frac{\partial^2 Q}{\partial \theta_M \partial \theta'_M}\right)\right\} < \infty,$$

and there is $\delta_M > 0$ such that

$$\sup_{|\tilde{\theta}_M - \theta_M| \leq \delta_M} \left|\frac{\partial^3 \tilde{Q}}{\partial \theta_{M,j} \partial \theta_{M,k} \partial \theta_{M,l}}\right| = O_\mathrm{P}(a_n), \quad 1 \leq j, k, l \leq p_M,$$

where $p_M = \dim(\theta_M)$.

Consider the fence inequality (1.20), where the constant c is, actually, dependent on the sample size. Thus, we can write $c = c_n$. Also recall the standard error $s(M, \tilde{M})$ in (1.20). Let $\sigma(M, \tilde{M})$ denote the standard deviation of $Q(M) - Q(\tilde{M})$ so that $s(M, \tilde{M})$ is an estimator of $\sigma(M, \tilde{M})$. We make the following additional assumptions.

A5. $c_n \to \infty$; and, for any true model M^* and incorrect model M, we have $\mathrm{E}\{Q(M)\} > \mathrm{E}\{Q(M^*)\}$, $\liminf\{\sigma(M, M^*)/a_n^{2\gamma-1}\} > 0$ and $c_n\sigma(M, M^*)/[\mathrm{E}\{Q(M)\} - \mathrm{E}\{Q(M^*)\}] \to 0$.

A6. $s(M, M^*) > 0$; $s(M, M^*) = \sigma(M, M^*)O_\mathrm{P}(1)$ if M^* is true and M is incorrect; and $\sigma(M, M^*) \vee a_n^{2\gamma-1} = s(M, M^*)O_\mathrm{P}(1)$ if both M and M^* are true models.

Note. *A1* is satisfied if $\mathrm{E}\{Q(M)\}$ can be differentiated under the expectation sign. *A2* implies that $|\hat{\theta}_M| \leq B_M$, where $\hat{\theta}_M$ is the minimizer of $Q(M, \theta_M)$ over $\theta_M \in \Theta_M$. To illustrate *A4* and *A5*, consider the case of clustered responses [see (1.15)]. Then, under regularity conditions, *A4* holds with $a_n = m$ and $\gamma = 1/2$. Furthermore, we have $\sigma(M, M^*) = O(\sqrt{m})$ and $\mathrm{E}\{Q(M)\} - \mathrm{E}\{Q(M^*)\} = O(m)$, provided that M^* is true, M is incorrect, and some regularity conditions hold. Thus, *A5* holds with $\gamma = 1/2$ and c_n being any sequence satisfying $c_n \to \infty$ and $c_n/\sqrt{m} \to 0$. Finally, *A6* does not require that $s(M, M^*)$ be a consistent estimator of $\sigma(M, M^*)$–only that it has the same order as the latter.

We first state a simple lemma that is used in the proof of our first consistency result.

Lemma 9.1. Under assumptions *A1–A4*, we have $\hat{\theta}_M - \theta_M = O_\mathrm{P}(a_n^{\gamma-1})$ and $Q(M, \hat{\theta}_M) - Q(M, \theta_M) = O_\mathrm{P}(a_n^{2\gamma-1})$ for every $M \in \mathcal{M}$.

Let M_0 be the model selected by fence using (1.20), where $c = c_n$ is considered a sequence of constants. The following theorem establishes consistency of the fence under this (non-adaptive) setting. The proofs of Lemma 9.1 and Theorem 9.1 can be found in Jiang *et al.* (2008).

Theorem 9.1. Under assumptions *A1–A6*, we have with probability tending to one that M_0 is a true model with minimum dimension.

Next, we consider consistency of the F-B fence, introduced in Section 1.5. The method was introduced in the case of extended GLMM, but the idea can be extended to other types of models. Let $M_0^\dagger$ denote the final model selected by the F-B fence procedure based on (1.20), where $c = c_n$ is a sequence of constants.

Theorem 9.2. Under assumptions *A1–A6*, we have with probability tending to one that $M_0^\dagger$ is a true model and no proper submodel of $M_0^\dagger$ is a true model.

The proof of Theorem 9.2 can be found in Jiang *et al.* (2008). Note that, here, consistency is in the sense that, with probability tending to one, $M_0^\dagger$ is a true model which cannot be further reduced or simplified. However, unlike the conclusion of Theorem 9.1, $M_0^\dagger$ is not, with probability tending to one, necessarily the optimal model, that is, a true model with the minimum dimension.

Finally, we give sufficient conditions for the consistency of the AF procedure, introduced in Section 3.1. For simplicity, assume that M_{opt} is unique. Consider the ratio

$$r_M = \frac{Q(M, \hat{\theta}_M) - Q(M_{\mathrm{f}}, \hat{\theta}_{M_{\mathrm{f}}})}{s(M, M_{\mathrm{f}})},$$

$M \in \mathcal{M}$. Let $\mathcal{M}_{\mathrm{w}\leq}$ denote the subset of incorrect models with dimension $\leq |M_{\mathrm{opt}}|$, where M_{opt} is the optimal model. Write $r_{\mathrm{opt}} = r_{M_{\mathrm{opt}}}$ and $r_{\mathrm{w}\leq} = \min_{M \in \mathcal{M}_{\mathrm{w}\leq}} r_M$. Denote the cumulative distribution functions (cdf's) of r_{opt} and $r_{\mathrm{w}\leq}$ by F_{opt} and $F_{\mathrm{w}\leq}$, respectively. Let $M_0(x)$ be the model selected by the fence procedure using (1.20) with $c = x$, and $P(x) = \mathrm{P}(M_0(x) = M_{\mathrm{opt}})$. Let $P^*(x)$ be the bootstrap version of $P(x)$. Denote the bootstrap sample size by B. Let a_n be a sequence of constants such that

$$\mathrm{E}\{Q(M, \hat{\theta}_M) - Q(M_{\mathrm{f}}, \hat{\theta}_{M_{\mathrm{f}}})\} \sim a_n.$$

Also let M_* denote the model with minimum dimension (assumed unique). We make the following additional assumptions.

A7. (Asymptotic distributional separation) if $M_{\mathrm{opt}} \notin \{M_{\mathrm{f}}, M_*\}$, then for any $\epsilon > 0$, there is $0 < \delta \leq 0.1$, $x_{n,1} < x_{n,2} < x_{n,3}$, and $N \geq 1$ such that when $n \geq N$ the following hold: $F_{\mathrm{opt}}(x_{n,1}) > 1 - \epsilon$, $F_{\mathrm{w}\leq}(x_{n,3}) \leq \epsilon$, $P(x_{n,2}) > 1 - \delta$, $1 - 4\delta < P(x_{n,j}) \leq 1 - 3\delta$, $j = 1, 3$.

A8. (Good bootstrap approximation) if $M_{\mathrm{opt}} \notin \{M_{\mathrm{f}}, M_*\}$, then for any $\delta, \eta > 0$, there are $N \geq 1$, $B^* = B^*(n)$ such that, when $n \geq N$ and $B \geq B^*$, we have $\mathrm{P}(\sup_{x>0} |P^*(x) - P(x)| < \delta) > 1 - \eta$.

For the most part, assumption *A7* says that there is an asymptotic separation between the optimal model and the incorrect ones that matter in that the peak of $P(x)$ is distant from the area where $r_{\mathrm{w}\leq}$ concentrates. This is reasonable because, typically, r_{opt} is of lower order than $r_{\mathrm{w}\leq}$. Therefore, one can find an interval, $(x_{n,1}, x_{n,3})$, such that (1.20) is almost always satisfied by $M = M_{\mathrm{opt}}$ when $c = c_n \in (x_{n,1}, x_{n,3})$. On the other hand, $(x_{n,1}, x_{n,3})$ is distant from the area where $r_{\mathrm{w}\leq}$ concentrates, so that $r_{\mathrm{opt}} \leq c_n$, $r_{\mathrm{w}\leq} > c_n$ with high probability, if $c_n \in (x_{n,1}, x_{n,3})$. Thus, $P(x)$ is expected to peak in $(x_{n,1}, x_{n,3})$ while $F_{\mathrm{w}\leq}(x)$ stays low in the region.

Recall that p^* in the AF procedure is a function of $c = c_n$, i.e., $p^* = p^*(c)$. Jiang *et al.* (2008) proved the following result that establishes consistency of the AF. For simplicity, here we confine to the case that the optimal model is neither the minimum model, M_*, nor the full model, M_f; however, the results of Jiang *et al.* (2008) also cover these two extreme cases.

Theorem 9.3. Suppose that assumptions *A7* and *A8* hold and $M_{opt} \notin \{M_f, M_*\}$. Then, the following hold:

(i) With probability tending to one there is $c^* \in (0, \infty)$ which is at least a local maximum and approximate global maximum of p^* in the sense that for any $\delta, \eta > 0$, there is $N \geq 1$ and $B^* = B^*(n)$ such that $P(p^*(c^*) \geq 1 - \delta) \geq 1 - \eta$, if $n \geq N$ and $B \geq B^*$.

(ii) Define c^* as the one in (i), if the latter exists, and $c^* = c_1$ otherwise, where c_1 is a fixed cut-off point. Let M_0^* be the model selected by the fence procedure using (1.20) with $\tilde{M} = M_f$, and c replaced by c^*. Then M_0^* is consistent in the sense that for any $\eta > 0$ there is $N \geq 1$ and $B^* = B^*(n)$ such that $P(M_0^* = M_{opt}) \geq 1 - \eta$, if $n \geq N$ and $B \geq B^*$.

9.3 Asymptotic properties of shrinkage selection

In this section, we consider asymptotic properties of the BS method, introduced in Section 7.1, for choosing the regularization parameters in the shrinkage selection methods. The first result is regarding consistency in model selection, that is, identification of the nonzero regression coefficients in the linear model, in which the number of potential predictors, p, may be (much) larger than the sample size, n. As noted (see Section 7.3), the BS method may be viewed as a special case of the AF, so a consistency result can be derived from the general result in the previous section. Nevertheless, Pang *et al.* (2014) proved the following result directly, without using the consistency of AF. The similarity to Theorem 9.3 is apparent.

We shall focus on the penalized likelihood method, (7.1), with the penalty $\rho(\lambda_j, |\beta_j|) = \lambda \rho_j(|\beta_j|)$, where $\rho_j(\cdot)$ is a known function. This includes Lasso and adaptive Lasso as special cases (see Section 7.1). Note that, in the latter case, one has $\rho_j(|\beta_j|) = |\beta_j|/|\hat{\beta}_j^{(0)}|$, where $\hat{\beta}_j^{(0)}$ is a shrinkage estimator of β_j with a given (small) $\lambda = \lambda_0$ [Zou (2006)]. For example, if $p < n$, the ordinary least squares (OLS)estimator, which corresponds to $\lambda_0 = 0$, may be used as $\hat{\beta}_j^{(0)}$. Let λ_M denote the regularization parameter, λ, such that, with this λ, the penalized likelihood method selects model M (if there are more than one such λ's, take the infimum). Write $\lambda_{opt} = \lambda_{M_{opt}}$, where M_{opt} is the optimal model corresponding to the non-zero regression coefficients (and nonthing else). Let $\lambda_\leq = \min_{M \in \mathcal{M}_\leq} \lambda_M$, where $\mathcal{M}$ denotes the subset of incorrect models with dimension $\leq |M_{\text{opt}}|$. Denote the cdf's of λ_{opt} and $\lambda_{\leq}$ by F_{opt} and $F_{\leq}$, respectively. Let M_λ be the selected model corresponding to the given λ, and $\Pr(\lambda) = P(M_\lambda = M_{\text{opt}})$. Let $\Pr^*(\lambda)$ be the bootstrap version of $\Pr(\lambda)$. Again, we use B to denote the bootstrap sample size, M_* for the minimum model (with all of the regression coefficients equal to zero), and M_{f} for the full model (with all of the regression coefficients non-zero). We assume that the following assumptions hold when $M_{\text{opt}} \notin \{M_*, M_{\text{f}}\}$, which are similar to assumptions $A7$ and $A8$ of the previous section.

$B7$. For any $\epsilon > 0$, there is $0 < \delta < 0.1$, $\lambda_{n,1} < \lambda_{n,2} < \lambda_{n,3}$, and $N \geq 1$ such that when $n \geq N$, the following hold: $F_{\text{opt}}(\lambda_{n,1}) > 1-\epsilon$, $F_{\leq}(\lambda_{n,3}) \leq \epsilon$, $\Pr(\lambda_{n,2}) > 1 - \delta$, and $1 - 4\delta < \Pr(\lambda_{n,j}) \leq 1 - 3\delta, j = 1, 3$.

$B8$. For any $\delta, \eta > 0$, there are $N \geq 1$ and $B^* = B^*(n)$ such that, when $n \geq N$ and $B \geq B^*$, we have $P\{|\Pr^*(\lambda_{n,j}) - \Pr(\lambda_{n,j})| < \delta\} > 1 - \eta, j = 1, 2, 3$, where the $\lambda_{n,j}$'s are as in $B7$.

As in Section 9.2, assumption $B7$ is related to "asymptotic distributional separation", and assumption $B8$ may be interpreted as "good bootstrap approximation". Recall that $p(\lambda)$ is the empirical (bootstrap) probability of the most frequently selected model, given λ.

Theorem 9.4. Suppose that $M_{\text{opt}} \notin \{M_*, M_{\text{f}}\}$ and assumptions $B7$, $B8$ hold. Then, the following hold.

(i) With probability tending to one, there is $\lambda^* \in (0, \infty)$, which is at least a local maximum and approximate global maximum of $p(\cdot)$, with the following property: for any $\eta, \delta > 0$, there is $N \geq 1$ and $B^* = B^*(n)$ such that $P\{p(\lambda^*) \geq 1 - \delta\} \geq 1 - \eta$, if $n \geq N$ and $B \geq B^*$.

(ii) Let M^* be the model selected via (7.1) with $\lambda = \lambda^*$. Then, M^* is consistent in the sense that for any $\eta > 0$, there is $N \geq 1$ and $B^* = B^*(n)$ such that $P(M^* = M_{\text{opt}}) \geq 1 - \eta$, if $n \geq N$ and $B \geq B^*$.

Below we illustrate the two assumptions of Theorem 9.4, $B7$ and $B8$, with an example. A modification of this example can be used to illustrate assumptions $A7$ and $A8$ in the previous section.

Example 9.1. Consider linear regression with two orthogonal predictors. The model can be expressed as

$$y_i = \beta_1 x_{i1} + \beta_2 x_{i2} + \epsilon_i, \tag{9.2}$$

$i = 1, \ldots, n$, where the ϵ_i's are i.i.d. $N(0, 1)$ (assume the error variance is known for simplicity); x_{i1}, x_{i2} satisfies $\sum_i x_{i1} = \sum_i x_{i2} = 0$, $\sum_i x_{i1}^2 = \sum_i x_{i2}^2 = n$, and $\sum_i x_{i1} x_{i2} = 0$. To see a special case, suppose that $n = 4k$ for some integer $k \geq 1$. Let $x_{i1} = 1, 1 \leq i \leq 2k$; $x_{i1} = -1, 2k+1 \leq i \leq 4k$;

$x_{i2} = (-1)^i, 1 \le i \le n$. The model can be written as $y = X\beta + \epsilon$, where $X'X = nI_2$ and I_2 is the 2×2 identity matrix. Write (9.2) as

$$y_i = b_1\tilde{x}_{i1} + b_2\tilde{x}_{i2} + \epsilon_i, \tag{9.3}$$

where $b_j = \sqrt{n}\beta_j$ and $\tilde{x}_{ij} = x_{ij}/\sqrt{n}$, $j = 1, 2$. Then, (9.3) can be written as $y = \tilde{X}b + \epsilon$, where $\tilde{X}'\tilde{X} = I_2$. The Lasso solution to (9.3) is

$$\hat{b}_j = \text{sign}(\hat{b}_j^0)\left(|\hat{b}_j^0| - \frac{\lambda}{2}\right)^+, \tag{9.4}$$

$j = 1, 2$, where $\hat{b}^0 = (\hat{b}_1^0, \hat{b}_2^0)'$ is the OLS estimator of $b = (b_0, b_1)'$, $x^+ = x \vee 0$, and $\text{sign}(x) = 1_{(x>0)} - 1_{(x<0)}$. Although (9.4) is a well-known result, the reader is invited to provide a simple proof of it (Exercise 9.1). Thus, equivalently, the Lasso solution to (9.2) is

$$\hat{\beta}_j = \text{sign}(\hat{\beta}_j^0)\left(\sqrt{n}|\hat{\beta}_j^0| - \frac{\lambda}{2}\right)^+, \tag{9.5}$$

$j = 1, 2$, where $\hat{\beta}^0 = (\hat{\beta}_1^0, \hat{\beta}_2^0)'$ is the OLS estimator of $\beta = (\beta_1, \beta_2)'$.

Suppose that the true parameters are $\beta_1 = 1$ and $\beta_2 = 0$. Then, by the consistency of the OLS estimator, we have

$$\text{P}(|\hat{\beta}_1^0| > |\hat{\beta}_2^0|) \longrightarrow 1, \tag{9.6}$$

as $n \to \infty$. If $|\hat{\beta}_1^0| > |\hat{\beta}_2^0|$, then it is easy to show that $M_\lambda = M_f$, the full model (with $\beta_1 \neq 0, \beta_2 \neq 0$), if $0 \le \lambda < 2\sqrt{n}|\hat{\beta}_2^0|$; $M_\lambda = M^*$, the optimal model (with $\beta_1 \neq 0, \beta_2 = 0$), if $2\sqrt{n}|\hat{\beta}_2^0| \le \lambda < 2\sqrt{n}|\hat{\beta}_1^0|$; and $M_\lambda = M_*$, the minimum model (with $\beta_1 = \beta_2 = 0$), if $\lambda \ge 2\sqrt{n}|\hat{\beta}_1^0|$ (Exercise 9.1). It follows that $\lambda_{\text{opt}} = 2\sqrt{n}|\hat{\beta}_2^0|$ and $\lambda_{\le} = 2\sqrt{n}|\hat{\beta}_1^0|$. Furthermore, it is easy to show that $\hat{\beta}_1^0, \hat{\beta}_2^0$ are independent with $\hat{\beta}_1^0 \sim N(1, n^{-1})$ and $\hat{\beta}_2^0 \sim N(0, n^{-1})$ (Exercise 9.1). Therefore, combining with (9.6), it is easy to show that

$$F_{\text{opt}}(\lambda) = 2\Phi\left(\frac{\lambda}{2}\right) - 1 + o(1), \tag{9.7}$$

$$F_{\le}(\lambda) = \Phi\left(\frac{\lambda}{2} - \sqrt{n}\right) + \Phi\left(\frac{\lambda}{2} + \sqrt{n}\right) - 1 + o(1), \tag{9.8}$$

$$\text{Pr}(\lambda) = F_{\text{opt}}(\lambda)\{1 - F_{\le}(\lambda)\} + o(1), \tag{9.9}$$

where $\Phi(\cdot)$ is the cdf of $N(0, 1)$, and the $o(1)$'s converge to zero as $n \to \infty$ uniformly for all λ (Exercise 9.2). Let $F_{\text{opt}}^0(\lambda)$, $F_{\le}^0(\lambda)$ and $P^0(\lambda)$ denote the right side of (9.7)–(9.9) without the $o(1)$, respectively.

For any $\epsilon > 0$, let $\delta = (\epsilon/7) \wedge 0.1$, $\Lambda_1 = 2\Phi^{-1}(1 - \epsilon/4)$, and $\Lambda_2 = 2\Phi^{-1}(1 - \delta/8)$. Then, we have $\Lambda_1 < \Lambda_2$, $F_{\text{opt}}^0(\Lambda_1) = 1 - \epsilon/2$, and $F_{\text{opt}}^0(\Lambda_2) =$

$1-\delta/4$. Note that $\Lambda_j, j = 1, 2$ do not depend on n. For any fixed λ, we have $F^0_{\leq}(\lambda) \to 0$ as $n \to \infty$. Thus, there is $N_1 \geq 1$ such that $F^0_{\leq}(\Lambda_2) < \delta/4, n \geq N_1$. On the other hand, for each (fixed) n such that $n \geq N_1$, $F^0_{\leq}(\cdot)$ is continuous and strictly increasing such that $F^0_{\leq}(\Lambda_2) < \delta/4$ and $F^0_{\leq}(\lambda) \to 1$ as $\lambda \to \infty$. Thus, there are $\Lambda_{n,4} > \Lambda_{n,3} > \Lambda_2$ such that $F^0_{\leq}(\Lambda_{n,3}) = \delta/4$ and $F^0_{\leq}(\Lambda_{n,4}) = \epsilon/2$. Thus, we have

$$P^0(\Lambda_1) < F^0_{\text{opt}}(\Lambda_1) = 1 - \epsilon/2 \leq 1 - (7/2)\delta,$$
$$P^0(\Lambda_{n,3}) = F^0_{\text{opt}}(\Lambda_{n,3})(1 - \delta/4) > F^0_{\text{opt}}(\Lambda_2)(1 - \delta/4)$$
$$= (1 - \delta/4)^2 > 1 - \delta/2,$$
$$P^0(\Lambda_{n,4}) < 1 - F^0_{\leq}(\Lambda_{n,4}) = 1 - \epsilon/2 \leq 1 - (7/2)\delta.$$

Let $\lambda_{n,2}$ be a point on $[\Lambda_1, \Lambda_{n,4}]$ so that $P^0(\cdot)$ attains its maximum over the interval at $\lambda_{n,2}$. It follows that $\lambda_{n,2} \in (\Lambda_1, \Lambda_{n,4})$ and $P^0(\lambda_{n,2}) \geq P^0(\Lambda_{n,3}) > 1 - \delta/2$. On the other hand, because $P^0(\Lambda_1) < 1 - (7/2)\delta < 1 - \delta/2$, there is $\lambda_{n,1} \in (\Lambda_1, \lambda_{n,2})$ such that $P^0(\lambda_{n,1}) = 1 - (7/2)\delta$. Similarly, there is $\lambda_{n,3} \in (\lambda_{n,2}, \Lambda_{n,4})$ such that $P^0(\lambda_{n,3}) = 1 - (7/2)\delta$.

In conclusion, there is $\delta \in (0, 0.1]$, $\lambda_{n,1} < \lambda_{n,2} < \lambda_{n,3}$, and $N_1 \geq 1$ such that, when $n \geq N_1$, we have

$$F^0_{\text{opt}}(\lambda_{n,1}) > F^0_{\text{opt}}(\Lambda_1) = 1 - \epsilon/2,$$
$$F^0_{\leq}(\lambda_{n,3}) < F^0_{\leq}(\Lambda_{n,4}) = \epsilon/2,$$
$$P^0(\lambda_{n,2}) > 1 - \delta/2, \quad \text{and}$$
$$P^0(\lambda_{n,1}) = P^0(\lambda_{n,3}) = 1 - (7/2)\delta.$$

On the other hand, there is $N_2 \geq 1$ such that, when $n \geq N_2$, all of the $o(1)$'s in (9.7)–(9.9) are less than $\delta/2 < \epsilon/2$ [see the note below (9.9)]. Therefore, when $n \geq N = N_1 \vee N_2$, we have, by (9.7)–(9.9),

$$F_{\text{opt}}(\lambda_{n,1}) > F^0_{\text{opt}}(\lambda_{n,1}) - \epsilon/2 > 1 - \epsilon,$$
$$F_{\leq}(\lambda_{n,3}) < F^0_{\leq}(\lambda_{n,3}) + \epsilon/2 < \epsilon,$$
$$P(\lambda_{n,2}) > P^0(\lambda_{n,2}) - \delta/2 > 1 - \delta,$$
$$P(\lambda_{n,j}) > P^0(\lambda_{n,j}) - \delta/2 = 1 - 4\delta, \quad \text{and}$$
$$P(\lambda_{n,j}) < P^0(\lambda_{n,j}) + \delta/2 = 1 - 3\delta, \quad j = 1, 3.$$

Therefore, asumption *B7* is verified.

Now consider assumption *B8*. Let $y_{(b)} = [y_i^{(b)}]_{1 \leq i \leq n}$ denote the bth bootstrap sample, $1 \leq b \leq n^*$. Then, we have $y^{(b)} = X\hat{\beta} + \epsilon^{(b)}$, or

$$y_i^{(b)} = \hat{\beta}_1 x_{i1} + \hat{\beta}_2 x_{i2} + \epsilon_i^{(b)},$$

$i = 1, \ldots, n$, where $\hat{\beta} = (\hat{\beta}_1, \hat{\beta}_2)'$ is the Lasso solution, based on the original data, under M_f, and $\epsilon^{(b)} = [\epsilon_i^{(b)}]_{1 \le i \le n} \sim N(0, I_n)$. By the same argument as above (see Exercise 9.1), it can be shown that, if $|\hat{\beta}_1^{0(b)}| > |\hat{\beta}_2^{0(b)}|$, where $\hat{\beta}^{0(b)} = [\hat{\beta}_1^{0(b)}, \hat{\beta}_2^{0(b)}]'$ is the OLS estimator based on $y^{(b)}$, then $M_\lambda^{(b)} = M^*$ iff $2\sqrt{n}|\hat{\beta}_2^{0(b)}| \le \lambda < 2\sqrt{n}|\hat{\beta}_1^{0(b)}|$, where $M_\lambda^{(b)}$ is the selected model based on $y^{(b)}$, for the given λ. Thus, we have

$$\text{Pr}^*(\lambda) = \frac{1}{B} \sum_{b=1}^{B} 1_{(2\sqrt{n}|\hat{\beta}_2^{0(b)}| \le \lambda < 2\sqrt{n}|\hat{\beta}_1^{0(b)}|)} + \zeta_{n,B}, \qquad (9.10)$$

where $|\zeta_{n,B}| \le B^{-1} \sum_{b=1}^{B} 1_{(|\hat{\beta}_1^{0(b)}| \le |\hat{\beta}_2^{0(b)}|)}$. On the other hand, it is easy to show (Exercise 9.3) that

$$\hat{\beta}_j^{0(b)} = \hat{\beta}_j^0 + \xi_{b,j}/\sqrt{n}, \quad j = 1, 2,$$

where $\xi_{b,j}, 1 \le b \le B, j = 1, 2$ are independent $N(0, 1)$ random variables. Thus, by (9.10), we have

$$\text{Pr}^*(\lambda) = \frac{1}{B} \sum_{b=1}^{B} \left\{ 1_{(2\sqrt{n}|\hat{\beta}_2^{0(b)}| \le \lambda)} - 1_{(2\sqrt{n}|\hat{\beta}_1^{0(b)}| \le \lambda)} 1_{(2\sqrt{n}|\hat{\beta}_2^{0(b)}| \le \lambda)} \right\} + \zeta_{n,B}$$

$$= \frac{1}{B} \sum_{b=1}^{B} \sum_{j=1}^{2} (-1)^{j-1} 1_{\{\xi_{b,2} \le (-1)^{j-1}\lambda/2 - \sqrt{n}\hat{\beta}_2^0\}}$$

$$- \frac{1}{B} \sum_{b=1}^{B} \left[\sum_{j=1}^{2} (-1)^{j-1} 1_{\{\xi_{b,1} \le (-1)^{j-1}\lambda/2 - \sqrt{n}\hat{\beta}_1^0\}} \right]$$

$$\times \left[\sum_{k=1}^{2} (-1)^{k-1} 1_{\{\xi_{b,2} \le (-1)^{k-1}\lambda/2 - \sqrt{n}\hat{\beta}_2^0\}} \right] + \zeta_{n,B}$$

$$= \sum_{j=1}^{2} (-1)^{j-1} \frac{1}{B} \sum_{b=1}^{B} 1_{\{\xi_{b,2} \le (-1)^{j-1}\lambda/2 - \sqrt{n}\hat{\beta}_2^0\}}$$

$$- \sum_{j,k=1}^{2} (-1)^{j+k} \frac{1}{B} \sum_{b=1}^{B} 1_{\{\xi_{b,1} \le (-1)^{j-1}\lambda/2 - \sqrt{n}\hat{\beta}_1^0, \xi_{b,2} \le (-1)^{k-1}\lambda/2 - \sqrt{n}\hat{\beta}_2^0\}}$$

$$+ \zeta_{n,B}. \qquad (9.11)$$

By the (multivariate) Glivenko-Cantelli theorem [e.g., Shorack and Wellner (1986), ch. 26), we have

$$\Delta_1 = \sup_x \left| \frac{1}{B} \sum_{b=1}^{B} 1_{(\xi_{b,2} \le x)} - \Phi(x) \right| \xrightarrow{\text{a.s.}} 0, \qquad (9.12)$$

$$\Delta_2 = \sup_{x_1, x_1} \left| \frac{1}{B} \sum_{b=1}^{B} 1_{(\xi_{b,1} \le x_1, \xi_{b,2} \le x_2)} - \Phi(x_1)\Phi(x_2) \right| \xrightarrow{\text{a.s.}} 0, \qquad (9.13)$$

as $B \to \infty$, where $\Phi(\cdot)$ is the cdf of $N(0,1)$ [note that $\xi_{b,1}, \xi_{b,2}$ are independent $N(0,1)$]. Thus, by (9.11), we have

$$\Pr^*(\lambda) = \sum_{j=1}^{2}(-1)^{\hat{\jmath}-1}\Phi\left\{(-1)^{j-1}\frac{\lambda}{2} - \sqrt{n}\hat{\beta}_2^0\right\}$$

$$- \sum_{j,k=1}^{2}(-1)^{j+k}\Phi\left\{(-1)^{j-1}\frac{\lambda}{2} - \sqrt{n}\hat{\beta}_1^0\right\}\Phi\left\{(-1)^{k-1}\frac{\lambda}{2} - \sqrt{n}\hat{\beta}_2^0\right\}$$

$$+ \eta_{n,B} + \zeta_{n,B}, \tag{9.14}$$

where $|\eta_{n,B}| \leq \Delta_1 + 4\Delta_2$.

Next, it is easy to show that $\sqrt{n}\hat{\beta}_j^0 = (2-j)\sqrt{n} + \xi_j, j = 1, 2$, where ξ_1, ξ_2 are independent $N(0,1)$ (Exercise 9.3). By Taylor expansion, we have

$$\sum_{j=1}^{2}(-1)^{j-1}\Phi\left\{(-1)^{j-1}\frac{\lambda}{2} - \sqrt{n}\hat{\beta}_2^0\right\}$$

$$= \sum_{j=1}^{2}(-1)^{j-1}\Phi\left\{(-1)^{j-1}\frac{\lambda}{2}\right\}$$

$$+ \xi_2 \sum_{j=1}^{2}(-1)^{j}\phi\left\{(-1)^{j-1}\frac{\lambda}{2} - \tilde{\xi}_{2,j}\right\}, \tag{9.15}$$

where $\phi(\cdot)$ is the pdf of $N(0,1)$, and $\tilde{\xi}_{2,j}$ satisfies $|\tilde{\xi}_{2,j}| \leq |\xi_2|$. By (9.7), it is seen that the first term on the right side of (9.15) is equal to $F_{\mathrm{opt}}(\lambda) + o(1)$, where the $o(1)$ is the same as that in (9.7). On the other hand, denoting the second term on the right side of (9.15) by w_2, it is easy to show that

$$|w_2| \leq |\xi_2| \sup_{x \geq \lambda/2 - |\xi_2|} \phi(x). \tag{9.16}$$

Similarly, it can be shown that

$$\sum_{j=1}^{2}(-1)^{j-1}\Phi\left\{(-1)^{j-1}\frac{\lambda}{2} - \sqrt{n}\hat{\beta}_1^0\right\}$$

$$= \sum_{j=1}^{2}(-1)^{j-1}\Phi\left\{(-1)^{j-1}\frac{\lambda}{2} - \sqrt{n}\right\}$$

$$+ \xi_1 \sum_{j=1}^{2}(-1)^{j}\phi\left\{(-1)^{j-1}\frac{\lambda}{2} - \sqrt{n} - \tilde{\xi}_{1,j}\right\}$$

$$= F_{\leq}(\lambda) + o(1) + w_1,$$

where the $o(1)$ is the same as in (9.8), and

$$|w_1| \leq |\xi_1| \left[\left\{ \sup_{x \leq -\lambda/2 - \sqrt{n} + |\xi_1|} \phi(x) \right\} \right.$$
$$\left. \vee \left\{ \sup_{x \leq \lambda/2 - \sqrt{n} + |\xi_1|} \phi(x) \right\} \right]. \qquad (9.17)$$

Thus, combined with the results of (9.9), (9.12)–(9.14), we obtain

$$\begin{aligned} \mathrm{Pr}^*(\lambda) &= F_{\mathrm{opt}}(\lambda) + o(1) + w_2 \\ &\quad - \{F_{\mathrm{opt}}(\lambda) + o(1) + w_2\}\{F_{\leq}(\lambda) + o(1) + w_1\} \\ &= \mathrm{Pr}(\lambda) + w_1 w_2 + w_1 O(1) + w_2 O(1) + o(1) \\ &\quad + \eta_{n,B} + \zeta_{n,B}, \end{aligned} \qquad (9.18)$$

where the $O(1)$'s are bounded, the $o(1)$ converges to zero uniformly in λ as $n \to \infty$, and the $O(1)$'s and $o(1)$ do not depend on B; futhermore, $\eta_{n,B}, \zeta_{n,B}$ are as in (9.14); and w_1, w_2 satisfy (9.16), (9.17).

For any $\delta, \eta > 0$, first choose $M_1 \geq 1, N_1$ such that the $O(1)$'s in (9.18) are bounded by M_1 when $n \geq N_1$. Next, we find M_2 such that $\mathrm{P}(|\xi_j| \leq M_2, j = 1, 2) > 1 - \eta/4$. Finally, we choose $\epsilon > 0$ such that

$$\left\{ \sup_{x \geq \Phi^{-1}(1-\epsilon) - M_2} \phi(x) \right\} \vee \left\{ \sup_{x \leq \Phi^{-1}(3\epsilon) + M_2} \phi(x) \right\}$$
$$< \left(\frac{\delta}{6M_1} \right) \wedge 1. \qquad (9.19)$$

For the ϵ in (9.19), by the verified B7, we find $\lambda_{n,j}, j = 1, 2, 3$ that satisfy the assumption for $n \geq N_2$. Also, we find N_3 such that, when $n \geq N_3$, the probability is $> 1 - \eta/4$ that all the $o(1)$'s in (9.7)–(9.9) and (9.18) are bounded in absolute value by $(\delta/6) \wedge \epsilon$. Thus, when $n \geq N_2 \vee N_3$, the probability is $> 1 - \eta/2$ that

$$2\Phi(\lambda_{n,1}/2) - 1 + \epsilon \geq 2\Phi(\lambda_{n,1}/2) - 1 + o(1) = F_{\mathrm{opt}}(\lambda_{n,1}) > 1 - \epsilon,$$

implying $\lambda_{n,1}/2 > \Phi^{-1}(1 - \epsilon)$, hence

$$\lambda_{n,j}/2 - |\xi_2| \geq \lambda_{n,1}/2 - M_2 > \Phi^{-1}(1 - \epsilon) - M_2, \quad j = 1, 2, 3;$$

also, we have $\Phi(\lambda_{n,3}/2 + \sqrt{n}) > \Phi(\lambda_{n,1}/2) > 1 - \epsilon$, hence

$$\begin{aligned} \epsilon &\geq F_{\leq}(\lambda_{n,3}) = \Phi(\lambda_{n,3}/2 - \sqrt{n}) + \Phi(\lambda_{n,3}/2 + \sqrt{n}) - 1 + o(1) \\ &> \Phi(\lambda_{n,3}/2 - \sqrt{n}) + 1 - \epsilon - 1 - \epsilon = \Phi(\lambda_{n,3}/2 - \sqrt{n}) - 2\epsilon, \end{aligned}$$

implying

$$-\lambda_{n,j} - \sqrt{n} + |\xi_1| < \lambda_{n,j}/2 - \sqrt{n} + |\xi_1|$$
$$\leq \lambda_{n,3}/2 - \sqrt{n} + M_2 < \Phi^{-1}(3\epsilon) + M_2, \quad j = 1,2,3.$$

Thus, by (9.16) and (9.17), the probability is $> 1 - \eta/2$ that all of the w's corresponding to $\lambda = \lambda_{n,j}, j = 1,2,3$ are $<$, in absolute value, by the right side of (9.19), if $n \geq N_2 \vee N_3$. It follows that, when $n \geq N_1 \vee N_2 \vee N_3$, the probability is $> 1 - \eta/2$ that

$$|w_1 w_2 + w_1 O(1) + w_2 O(1) + o(1)|$$
$$< \delta/6 + (\delta/6M_1)M_1 + (\delta/6M_1)M_1 + \delta/6$$
$$= 2\delta/3.$$

Also, by (9.12) and (9.13), there is $B^* \geq 1$ such that, when $B \geq B^*$, the probability is $> 1 - \eta/4$ that $|\eta_{n,B}| < \delta/6$ [see the inequality below (9.14)].

Finally, because

$$E(|\zeta_{n,B}|) \leq P(|\hat{\beta}_1^{0(1)}| \leq |\hat{\beta}_2^{0(1)}|)$$
$$= P\{|1 + (\xi_1 + \xi_{1,1})/\sqrt{n}| \leq |(\xi_2 + \xi_{1,2})/\sqrt{n}|\} \to 0,$$

as $n \to \infty$ [see the expressions below (9.10), (9.14)], there is N_4 such that, when $n \geq N_4$, the probability is $> 1 - \eta/4$ that $|\zeta_{n,B}| < \delta/6$.

Therefore, in conclusion, when $n \geq N = N_1 \vee N_2 \vee N_3 \vee N_4$, the probability is $> 1 - \eta$ that $|\mathrm{Pr}^*(\lambda_{n,j}) - \mathrm{Pr}(\lambda_{n,j})| < 2\delta/3 + \delta/6 + \delta/6 = \delta$, $j = 1,2,3$. Assumption *B8* is thus verified because

$$P\{|\mathrm{Pr}^*(\lambda_{n,j}) - \mathrm{Pr}(\lambda_{n,j})| < \delta\} \geq P\left\{\max_{j=1,2,3} |\mathrm{Pr}^*(\lambda_{n,j}) - \mathrm{Pr}(\lambda_{n,j})| < \delta\right\}$$
$$> 1 - \eta, \quad j = 1,2,3.$$

The BS method is based on fitting the full model, and bootstrapping under the fitted model. This may not be possible if p is larger than n. In such a case, Pang *et al.* (2014) suggests to first apply the SIS method of Fan and Lv (2008) to reduce the dimension of the model to less than n, and then apply the BS method. Asymptotic property of such a combined procedure was studied by Pang *et al.* (2014), who proved that, under assumptions *B7* and *B8* and, in addition, conditions 1–4 of Theorem 1 of Fan and Lv (2008), the probability of selecting the optimal model goes to one, provided that $p > n$ and $\log(p) = O(n^\xi)$, where $\xi \in (0, 1 - 2\kappa)$ and κ satisfies

$$\min_{j \in M_{\mathrm{opt}}} |\beta_j| \geq \frac{c_1}{n^\kappa} \quad \text{and} \quad \min_{j \in M_{\mathrm{opt}}} |\mathrm{cov}(\beta_j^{-1} y_i, x_{ij})| \geq c_2 \qquad (9.20)$$

for some constants $c_1, c_2 > 0$, where y_i and x_{ij} are the ith components of y and x_j, respectively. Note that, here, the predictors $x_1, \ldots, x_p$ are considered as random vectors, and M_{opt} is understood as the subset of $\{1, \ldots, p\}$ such that $\beta_j \neq 0$ iff $j \in M_{\text{opt}}$. The latter is, of course, just another expression of the optimal model.

9.4   Signal consistency of IF

We proceed under the basic assumptions of Sections 5.1 and 5.2. In particular, we assume that the measure of lack-of-fit, Q, is subtractive, hence (5.1) holds for some s_j's, which we call scores. More specifically, we assume that the data can be expressed as

$$y_{il} = \begin{cases} \mu_{i1} + \epsilon_{il}, & 1 \leq l \leq n_1, \\ \mu_{i2} + \epsilon_{il}, & n_1 + 1 \leq l \leq n, \end{cases} \tag{9.21}$$

$1 \leq i \leq N$, where N may be the total number of genes; $1 \leq l \leq n_1$ and $n_1 + 1 \leq l \leq n$ correspond to the control and treatment, respectively; μ_{ik}, $k = 1, 2$ are the means of the controls and treatments, respectively; and n is the sample size (e.g., the number of microarrays). Let $n_2 = n - n_1$. Furthermore, we assume that the score s_j has the expression

$$s_j = \psi_j(\delta_j, \xi_j), \tag{9.22}$$

$j = 1, \ldots, m$, where m may be the total number of (candidate) gene-sets; δ_j is an unknown parameter; ξ_j is a vector of random variables that does not depend on δ_j; and $\psi_j(\cdot, \cdot)$ is a function. Let $M_0 = \{1 \leq j \leq m : \delta_j \neq 0\}$. For example, in gene set analysis, the gene-sets in M_0 are called differentially expressed (d.e.). We consider a specific example.

Example 9.2. Suppose that there are m gene-sets each with a single gene (so that $N = m$). The gene-set scores are the two-sample t-statistics, that is,

$$s_j = \frac{\bar{y}_{j2} - \bar{y}_{j1}}{s_{j,\text{p},y}\sqrt{n_1^{-1} + n_2^{-1}}} \quad \text{with} \quad s_{j,\text{p},y}^2 = \frac{(n_1 - 1)s_{j1}^2 + (n_2 - 1)s_{j2}^2}{n - 2},$$

where $\bar{y}_{j1} = n_1^{-1} \sum_{l=1}^{n_1} y_{jl}$, $\bar{y}_{j2} = n_2^{-1} \sum_{l=n_1+1}^{n} y_{jl}$,

$$s_{j1}^2 = \frac{1}{n_1 - 1} \sum_{l=1}^{n_1} (y_{jl} - \bar{y}_{j1})^2, \quad s_{j2}^2 = \frac{1}{n_2 - 1} \sum_{l=n_1+1}^{n} (y_{jl} - \bar{y}_{j2})^2.$$

It follows that

$$s_j = \frac{\delta_j + \bar{\epsilon}_{j2} - \bar{\epsilon}_{j1}}{s_{j,\text{p},\epsilon}(n_1^{-1} + n_2^{-1})^{1/2}},$$

which is in the form of (9.22) with $\delta_j = \mu_{j2} - \mu_{j1}$, $\xi_j = (\xi_{j1}, \xi_{j2})'$, where $\xi_{j1} = \bar{\epsilon}_{j2} - \bar{\epsilon}_{j1}$, $\xi_{j2} = s_{j,p,\epsilon}\sqrt{n_1^{-1} + n_2^{-1}}$, and $\psi_j(u, v) = (u + v_1)/v_2$ for $v = (v_1, v_2)'$. Note that in this case ψ_j does not depend on j.

Without loss of generality, assume that all the δ_j's are nonnegative. By signal consistency, we mean that, as $\Delta = \min_{j \in M_0} \delta_j \to \infty$, the probability of identifying (exactly) M_0 as the d.e. gene-sets goes to one. Note that, although we use, from time to time, the terms in gene set analysis, the concept can be extended to other cases as well. Also note that, although the sample size n is not required to go to infinity, it is necessary that the total number of observations, that is, Nn, goes to infinity, as $\Delta \to \infty$, in order to have signal consistency. We illustrate this with a simple example.

Example 9.3. (A counter example) Suppose that there are three gene-sets, each consists of a single gene. Let $l = 1, 2$ be the controls and $l = 3, 4$ the treatments. Let y_{jl}, $j = 1, 2, 3$, $l = 1, 2, 3, 4$ be the data so that $y_{jl} = \epsilon_{jl}$ for all j, l except for $j = 1$ and $l = 3, 4$, and $y_{1l} = \Delta + \epsilon_{1l}$, $l = 3, 4$, where $\Delta > 0$ and ϵ_{jl}'s are independent and distributed as $N(0, 1)$. The gene-set scores are defined as

$$s_j = \frac{y_{j3} + y_{j4}}{2} - \frac{y_{j1} + y_{j2}}{2}, \quad j = 1, 2, 3.$$

It is easy to show that s_j can be expressed as (9.22) (Exercise 9.4) with $\delta_1 = \Delta$, $\delta_2 = \delta_3 = 0$, ξ_j equal to s_j with y replaced by ϵ, $\psi_j(u, v) = u + v$, and $M_0 = \{1\}$. First note that there is a positive probability, say, p_0, such that 1) all of the ϵ_{il}'s are bounded in absolute value by one; 2) $\min_{l=3,4} \epsilon_{2l} > \max_{l=3,4} \epsilon_{3l}$; 3) $\max_{l=1,2} \epsilon_{2l} < \min_{l=1,2} \epsilon_{3l}$. Let $y_l = (y_{jl})_{j=1,2,3}$, $l = 1, 2, 3, 4$. Then, the bootstrap procedure draws samples, with replacements, y_1^*, y_2^* from $\{y_1, y_2\}$, and y_3^*, y_4^* from $\{y_3, y_4\}$. It is easy to see that, under 1)–3), no matter what the bootstrap samples, one always have $s_1^* > s_2^* \vee s_3^*$ and $s_2^* > s_3^*$, if $\Delta > 4$ (Exercise 9.4). There-fore, at dimension two one always selects the gene-sets 1 and 2, hence the corresponding p^* is 1. It follows, by the conservative principle, that the probability is at least p_0 that two gene-sets will be selected (note that, here, the full dimension, which is 3, is not considered in the IF, because the corresponding p^* is always one), no matter how large Δ is. It follows that the IF procedure is not signal consistent in this case.

In the following we assume that the IF frequencies, p^*, are compared for all of the dimensions $1 \le d \le K$, where $K < m$ and increases with m. This is practical because, in practice, one may not know an upper bound of the number of d.e. gene-sets. By letting K increase with m it guarantees that the range of d eventually covers $d_0 = |M_0|$. The rate at which K increases with m is subject to the constrains of the following assumptions. Let $\epsilon_l = (\epsilon_{il})_{1 \leq i \leq N}$, $1 \leq l \leq n$. Let $\hat{P}$ denote the "empirical distribution" of ϵ_l, $l = 1, \ldots, n$ that puts the mass $1/n_1^{n_1} n_2^{n_2}$ on every point in $x = (x_1, \ldots, x_n) \in R^{Nn}$, where $x_l = (x_{il})_{1 \leq i \leq N}$, such that $x_1, \ldots, x_{n_1} \in \{\epsilon_1, \ldots, \epsilon_{n_1}\}$ and $x_{n_1+1}, \ldots, x_n \in \{\epsilon_{n_1+1}, \ldots, \epsilon_n\}$. Note that this is not necessarily an empirical distribution in the usual sense [e.g., Shorack and Wellner (1986)], because the ϵ_l's may not be identically distributed, or even independent. Here come the key assumptions.

C1. (d.e. gene-sets) $d_0 > 0$; the probability $\to 1$, as $\Delta \to \infty$, that

$$\hat{P}\left(\min_{j \in M_0} s_j > \max_{j \notin M_0} s_j\right) = 1. \tag{9.23}$$

C2. (non-d.e. gene-sets) The probability $\to 1$ that

$$\hat{P}\left(\min_{j \in M} s_j \geq \max_{j \in M_1 \setminus M} s_j\right) < 1 \tag{9.24}$$

for every $M \subset M_1 = \{1, \ldots, m\} \setminus M_0$ with $1 \leq |M| \leq K - d_0$.

The first part of C1 states that there is at least one d.e. gene-set. To see what the second part of C1 means, let $\epsilon^{(b)}$ denote an arbitrary bootstrap sample from the sampling space above (that consists of $n_1^{n_1} n_2^{n_2}$ points in R^{Nn}). Let $y_l^{(b)} = \mu_1 + \epsilon_l^{(b)}$, $1 \leq l \leq n_1$, and $y_l^{(b)} = \mu_2 + \epsilon_l^{(b)}$, $n_1 + 1 \leq l \leq n$, where $\mu_k = (\mu_{ik})_{1 \leq i \leq N}$, $k = 1, 2$. Let $s_j^{(b)}$ be the corresponding s_j, $1 \leq j \leq m$. The notation $\forall b$ means for all possible bootstrap samples. To be more specific, we now focus on a special case.

Example 9.2 (continued). Suppose that the ϵ_{jl}'s are independent and distributed as $N(0, 1)$. Redefine the value of the denominator of s_j as 1 if it is equal to zero (this change has no practical impact since the probability that the denominator is zero is 0). Let $U_1 = \max_{\forall b} \max_{1 \leq j \leq m} |\bar{\epsilon}_{j2}^{(b)} - \bar{\epsilon}_{j1}^{(b)}|$, $U_2 = (n_1^{-1} + n_2^{-1}) \max_{\forall b} \max_{1 \leq j \leq m} s_{j,p,\epsilon^{(b)}}$, and U_3 be the minimum nonzero value of $(n_1^{-1} + n_2^{-1}) s_{j,p,\epsilon^{(b)}}$, $1 \leq j \leq m$, $\forall b$. For any $\rho > 0$, find $\lambda_k > 0$, $k = 1, 2, 3$ such that $P(U_k \leq \lambda_k, k = 1, 2, U_3 \geq \lambda_3) > 1 - \rho$. Let $\Delta_\rho = \lambda_1(1 + 1 \vee \lambda_2/1 \wedge \lambda_3)$. It can be shown (Exercise 9.5) that $\Delta > \Delta_\rho$, $U_k \leq \lambda_k$, $k = 1, 2$ and $U_3 \geq \lambda_3$ imply $s_j^{(b)} > \lambda_1/1 \wedge \lambda_3$, $j \in M_0$ and $s_j^{(b)} \leq \lambda_1/1 \wedge \lambda_3$, $j \notin M_0$ (regardless whether the denominator is zero), $\forall b$. Therefore, the probability is at least $1 - \rho$ that (9.23) holds, if $\Delta > \Delta_\rho$.

Note that $\Delta \to \infty$ is involved in C1 but not in C2, so it may be wondered what is the limiting process in C2. Recall the counterexample above (Example 9.3) showing that the size of the data, Nn, has to increase in order to have signal consistency. This is the limiting process implicitly assumed here. We illustrate *C2* with a one-sample problem for the sake of simplicity (the basic arguments for the two-sample problem are very similar).

Example 9.4. Suppose that the scores can be expressed as $s_j = \Delta 1_{(j=0)} + n^{-1} \sum_{l=1}^{n} \epsilon_{jl}$, $0 \leq j \leq m$, where $\Delta > 0$ and ϵ_{jl}'s are i.i.d. with a continuous distribution. Thus, $M_0 = \{0\}$ in this case (Exercise 9.6). Let A denote the complement of the event in *C2*, and A_M the complement of (9.24). Then, $A = \cup_{k=1}^{K-1} \cup_{M \subset \{1,\dots,m\}, |M|=k} A_M$. Note that

$$A_M = \left\{ \min_{i \in M} s_i^{(b)} \geq \max_{i \in \{1,\dots,m\} \setminus M} s_i^{(b)}, \ \forall b \right\}$$

for every $M \subset \{1,\dots,m\}$ (note that here the indexes start from 0). Let B (B_M) denote A (A_M) with $\geq$ replaced by $>$. It can be shown that

$$B_M = \{ \min_{i \in M} \epsilon_{il} > \max_{i \in \{1,\dots,m\} \setminus M} \epsilon_{il}, 1 \leq l \leq n \},$$

$\forall M \subset \{1,\dots,m\}$. Furthermore, given $1 \leq k \leq K$, the sets B_M for all $M \subset \{1,\dots,m\}$ with $|M| = k$ are disjoint with the same probability

$$\mathrm{P} \left(\min_{1 \leq j \leq k} \epsilon_{jl} > \max_{k+1 \leq j \leq m} \epsilon_{jl}, \ 1 \leq l \leq n \right)$$

$$= \prod_{l=1}^{n} \mathrm{P} \left(\min_{1 \leq j \leq k} \epsilon_{jl} > \max_{k+1 \leq j \leq m} \epsilon_{jl} \right) = \{C_k^m\}^{-n},$$

where C_k^m is the binomial coefficient. It follows that $\mathrm{P}(A) = \mathrm{P}(B) \leq$

$$\sum_{k=1}^{K-1} \sum_{M \subset \{1,\dots,m\}, |M|=k} \{C_k^m\}^{-n} = \sum_{k=1}^{K-1} \{C_k^m\}^{1-n}, \tag{9.25}$$

which goes to zero if either m, or n, increases, and K increases sufficiently slowly (Exercise 9.6).

The following result regarding consistency of the IF procedure was proved in Jiang *et al.* (2011b).

Theorem 9.5. Under assumptions *C1* and *C2*, we have with probability tending to one, as $\Delta, B \to \infty$, that
(i) $p^* = 1$ for $M = M_0$; and
(ii) $p^* < 1$ for all M such that $d_0 < |M| \leq K$;
hence, by the conservative principle, IF chooses M_0 as the d.e. gene-sets.

9.5 Convergence and consistency of E-MS with fence

To be specific, assume that the minimum-dimension criterion is used to select the model within the fence [see the paragraph after the one involving

(1.12) regarding the criterion used to select the model within the fence]. Again, assume the existence of $M_{\text{opt}} \in \mathcal{M}$. Also assume that model-based (or parametric) bootstrap is used in the AF procedure, with the bootstrap sample size B. Let $M^{(t)}$ be the current model. We assume that, given a model, the method of parameter estimation is determined (e.g., maximum likelihood with the E-M algorithm), so that we can simply use $P^*(\cdot|M)$ for the bootstrap empirical probability under model M. To be specific, let us focus on the fence based on (3.2). The discussion of this section also applies to the fence based on (1.20) with only minor changes.

Let $M^{(t)}$ be the current model. Then, for any cut-off c among a grid of values, $\mathcal{C}$, the updated model, $M^{(t+1)}$, is such that, for some $c^* \in \mathcal{C}$,

$$P^*\{M_{c^*} = M^{(t+1)}|M^{(t)}\} = \max_{M \in \mathcal{M}} P^*\{M_{c^*} = M|M^{(t)}\}$$

$$= \max_{c \in \mathcal{C}} \max_{M \in \mathcal{M}} P^*\{M_c = M|M^{(t)}\},$$

where M_c is the model selected by the fence for the cut-off c. Note that $\mathcal{C}$ is usually chosen so that the fence does not yield a trivial solution, that is, the minimum model, M_*, or the full model, M_{f}. The sequence of models $M^{(t)}, t = 0, 1, 2, \ldots$ is said to converge within $t + 1$ iterations if $M^{(t+1)} = M^{(t)}$. The following theorem assumes consistency of the AF, for which sufficient conditions are given in Section 9.2. The following result is due to Jiang *et al.* (2014) (Supplementary Material). As the proof is fairly simple, it is provided here for convenience.

Theorem 9.6. Provided that the AF is consistent when bootstrapping under either M_{f} or M_{opt}, as $n, B \to \infty$, then, with probability tending to one as $n, B \to \infty$, E-MS with AF converges within two iterations when starting with M_{f}. Furthermore, the limit of the convergence is M_{opt}; in other words, the E-MS with AF is consistent.

A nice feature of Theorem 9.6 is that it links the convergence of E-MS with AF to the consistency of AF, and shows that the convergence can be very fast (in two iterations) with the limit being the optimal model. In simulation study [Jiang *et al.* (2014), Subsection A.6.1], the E-MS with AF converged in 2-3 iterations in all of the simulation runs. On the other hand, the convergence is not global [e.g., Luenberger (1984)], because the starting model is assumed to be M_{f}. In fact, as is seen in the proof below, the starting model may be replaced by any overfitting model, but not by an arbitrary one.

Proof: Let $\mathcal{P}$ denote the joint probability distribution of the data and bootstrap samples. Then, we have

$$\mathcal{P}\{M^{(1)} = M_{\text{opt}}|M^{(0)} = M_{\text{f}}\} \to 1,$$

as $n, B \to \infty$. Also note that, given $M^{(1)}$, the outcome of $M^{(2)}$ does not depend on $M^{(0)}$. Thus, we have

$$\mathcal{P}\{M^{(2)} = M_{\text{opt}} | M^{(1)} = M_{\text{opt}}, M^{(0)} = M_{\text{f}}\}$$
$$= \mathcal{P}\{M^{(2)} = M_{\text{opt}} | M^{(1)} = M_{\text{opt}}\}$$
$$\to 1,$$

as $n, B \to \infty$. Therefore, we have

$$\mathcal{P}\{\text{E} - \text{MS converges in 2 iterations} | M^{(0)} = M_{\text{f}}\}$$
$$\geq \mathcal{P}\{M^{(2)} = M^{(1)} | M^{(0)} = M_{\text{f}}\}$$
$$\geq \mathcal{P}\{M^{(2)} = M_{\text{opt}}, M^{(1)} = M_{\text{opt}} | M^{(0)} = M_{\text{f}}\}$$
$$= \mathcal{P}\{M^{(1)} = M_{\text{opt}} | M^{(0)} = M_{\text{f}}\} \mathcal{P}\{M^{(2)} = M_{\text{opt}} | M^{(1)} = M_{\text{opt}}, M^{(0)} = M_{\text{f}}\}$$
$$\to 1, \quad \text{as } n, B \to \infty.$$

Also note that the E-MS stops whenever $M^{(t+1)} = M^{(t)}$ takes place, in which case the $M^{(t+1)}$ is the limit of convergence. Thus, we have

$$\mathcal{P}\{\text{the E} - \text{MS limit is } M_{\text{opt}} | M^{(0)} = M_{\text{f}}\}$$
$$\geq \mathcal{P}\{M^{(2)} = M_{\text{opt}}, M^{(1)} = M_{\text{opt}} | M^{(0)} = M_{\text{f}}\}$$
$$\to 1,$$

as $n, B \to \infty$, according to the previous argument, if $M_{\text{f}} \neq M_{\text{opt}}$. On the other hand, if $M_{\text{f}} = M_{\text{opt}}$, then, again by the previous argument, we have

$$\mathcal{P}\{\text{the E} - \text{MS limit is } M_{\text{opt}} | M^{(0)} = M_{\text{f}}\}$$
$$\geq \mathcal{P}\{M^{(1)} = M_{\text{opt}} | M^{(0)} = M_{\text{opt}}\}$$
$$\to 1, \quad \text{as } n, B \to \infty.$$

This proves the consistency.

A modification of the E-MS with AF, however, can actually achieve the global convergence, that is, the algorithm converges with any starting model, $M^{(0)}$. This is done by restricting the model space for $M^{(t+1)}$ to be submodels of $M^{(t)}$. This is not unreasonable because the bootstrap samples are drawn under $M^{(t)}$, which would suggest that $M^{(t)}$ is believed to be a true model; therefore, there is no need to look for anything beyond the submodels of $M^{(t)}$. The modified E-MS with AF is also computationally more attractive, because the model space shrinks with each iteration. As for the convergence property, we have the following result. The proof is left as an exercise (Exercise 9.7).

Theorem 9.7. If $\mathcal{M}$ is finite, then the modified E-MS with AF converges globally.

9.6   Concluding remark

In delivering his 2013 R. A. Fisher Lecture, Peter Bickel offered what he called a "humble view of future" for the Role of Statistician:

> 1. Applied statistics must engage theory and conversely.
> 2. More significantly it should be done in full ongoing collaboration with scientists and participation in the gathering of data and full knowledge of their nature.

As noted, the fence allows flexibility of choosing the optimal model within the fence. Namely, the criterion of optimality for the selection within the fence does not have to be statistical; it can incorporate scientific, economical, political, or even legal concerns. However, so far this potentially important aspect of the fence has yet to be explored [with, perhaps, the exception of Jiang *et al.* (2010)]. Of course, there may be a concern that, with a "right" choice of the optimality measure, a practitioner can always "legitimize" the selection in his/her favor, under the fence framework. Nevertheless, it is our strong belief that the practical problem, not the statistical one, is the only gold standard. The practical problem could be scientific, economical, political, legal, or of other natures. As long as the practioner is faithful to the best interest of his/her problem, the practioner can surely legitimize his or her choice. Therefore, in view of Professor Bickel's note, we expect this remarkable feature of the fence to be fully explored in the near future.

The future of the fence looks promising.

9.7   Exercises

9.1. This exercise is related to Example 9.1.

(i) Show that the Lasso solution to (9.3) is given by (9.4).

(ii) Suppose that $|\hat{\beta}_1^0| > |\hat{\beta}_2^0|$. Show that $M_\lambda = M_f$, the full model (with $\beta_1 \neq 0, \beta_2 \neq 0$), if $0 \leq \lambda < 2\sqrt{n}|\hat{\beta}_2^0|$; $M_\lambda = M^*$, the optimal model (with $\beta_1 \neq 0, \beta_2 = 0$), if $2\sqrt{n}|\hat{\beta}_2^0| \leq \lambda < 2\sqrt{n}|\hat{\beta}_1^0|$; and $M_\lambda = M_*$, the minimum model (with $\beta_1 = \beta_2 = 0$), if $\lambda \geq 2\sqrt{n}|\hat{\beta}_1^0|$.

(iii) Show that $\hat{\beta}_1^0$ and $\hat{\beta}_2^0$ are independent with the distributions $N(1, n^{-1})$ and $N(0, n^{-1})$, respectively.

9.2. Continuing from the previous exercise, establish (9.7)–(9.9).

9.3. Again, this exercise refers to Example 9.1, after (9.10).

(i) Show that $\hat{\beta}_j^{0(b)} = \hat{\beta}_j^0 + \xi_{b,j}/\sqrt{n}, j = 1, 2$, where $\xi_{b,j}, 1 \leq b \leq B, j =$

$1, 2$ are independent $N(0, 1)$ random variables.

(ii) Show that $\sqrt{n}\hat{\beta}_j^0 = (2 - j)\sqrt{n} + \xi_j, j = 1, 2$, where ξ_1, ξ_2 are independent $N(0, 1)$ random variables.

(iii) Verify (9.16) and (9.17).

9.4. This exercise is related to Example 9.3.

(i) Show that the gene-set score s_j defined therein can be expressed as (9.22) with $\delta_1 = \Delta$, $\delta_2 = \delta_3 = 0$, ξ_j equal to s_j with y replaced by ϵ, $\psi_j(u, v) = u + v$.

(ii) Show that $M_0 = \{1\}$.

(iii) Assuming that properties 1)–3) hold, show that, for any bootstrap samples, one has $s_1^* > s_2^* \vee s_3^*$ and $s_2^* > s_3^*$, if $\Delta > 4$.

9.5. In Example 9.2 (continued) following (9.24), show that $\Delta > \Delta_\rho$, $U_k \leq \lambda_k$, $k = 1, 2$ and $U_3 \geq \lambda_3$ imply $s_j^{(b)} > \lambda_1/1 \wedge \lambda_3$, $j \in M_0$ and $s_j^{(b)} \leq \lambda_1/1 \wedge \lambda_3$, $j \notin M_0$ (regardless whether the denominator is zero), $\forall b$.

9.6. This exercise is related to Example 9.4.

(i) Show that $M_0 = \{0\}$ in this case.

(ii) Show that the last expression of (9.25) goes to zero if either m, or n, increases, and K increases sufficiently slowly, and determine the rate of the latter increase of K.

9.7. Write a simple proof for Theorem 9.7.

Bibliography

Afifi, A. and Elashoff, R. (1966), *Missing observations in multivariate statistics: I. Review of the literature*, in *J. Amer. Statist. Assoc.* 61, pp. 595–604.

Akaike, H. (1973), *Information theory as an extension of the maximum likelihood principle*, in *Second International Symposium on Information Theory* (B. N. Petrov and F. Csaki eds.), pp. 267–281 (Akademiai Kiado, Budapest).

Allison, D. B., Fernandez, J. R., Heo, M. and Beasley, T. M. (2000), *Testing the robustness of the new Haseman-Elston quantitative-trait loci-mapping procedure*, in *Am. J. Hum. Genet.* 67, pp. 249–252.

Anderson, J. J. B. and Garner, S. C., eds (1995), *Calcium and Phosphorus in Health and Disease*, in *CRC Press* (Boca Raton, FL).

Akaike, H. (1974), *A new look at the statistical model identification*, in *IEEE Trans. on Automatic Control* Vol. 19, pp. 716–723.

Almasy, L. and Blangero, J. (1998), *Multipoint quantitative-trait linkage analysis in general pedigrees*, in *Am. J. Hum. Genet.* 62, 1198-1211.

Baierl, A., Bogdan, M., Frommlet, F., and Futschik, A. (2006), *On locating multiple interacting quantitative trait loci in intercross design*, in *Genetics* 173, pp. 1693–1703.

Ball, R. (2001), *Bayesian methods for quantitative trait loci mapping based on model selection: approximate analysis using the Bayesian information criterion*, in *Genetics* 159, pp. 1351–1364.

Battese, G. E., Harter, R. M., and Fuller, W. A. (1988), *An error-components model for prediction of county crop areas using survey and satellite data*, in *J. Amer. Statist. Assoc.* 80, pp. 28–36.

Beran, R. (1986), *Discussion of "Jackknife, bootstrap and other resampling methods in regression analysis" by C.F.J.Wu*, in *Annals of Statistics* 14, 1295-1298.

Bogdan, M., Ghosh, J. K., and Doerge, R. W. (2004), *Modifying the Schwarz Bayesian information criterion to locate multiple interacting quantitative trait loci*, in *Genetics* 167, pp. 989–999.

Bondell, H. D., Krishna, A. and Ghosh, S. K. (2010), *Joint variable selection for fixed and random effects in linear mixed-effects models*, in *Biometrics* 66, pp. 1069–1077.

Booth, J. G. and Hobert, J. P. (1999), *Maximizing generalized linear mixed model likelihoods with an automated Monte Carlo EM algorithm*, in *J. Roy. Statist. Soc. B* 61, pp. 265–285.

Bozdogan, H. (1987), *Model selection and Akaike's information criterion (AIC): the general theory and its analytical extensions*, in *Psychomatrika* 52, pp. 345–370.

Bozdogan, H. (1994), Editor's general preface, in *Engineering and Scientific Applications*, Vol. 3 (H. Bozdogan ed.), *Proceedings of the First US/Japan Conference on the Frontiers of Statistical Modeling: An Informational Approach*, pp. ix-xii (Kluwer Academic Publishers, Dordrecht, Netherlands).

Breiman, L., Friedman, J., Stone, C. J., and Olshen, R. A. (1984), *Classification and Regression Trees*, (Chapman & Hall/CRC, New York).

Breslow, N. E. and Clayton, D. G. (1993), *Approximate inference in generalized linear mixed models*, in *J. Amer. Statist. Assoc.* 88, pp. 9–25.

Broman, K. W. (1997), *Identifying quantitative trait loci in experimental crosses*, Ph.D. Dissertation, Univ. Calif., Berkeley, CA.

Broman, K. W. and Speed, T. P. (2002), *A model selection approach for the identification of quantitative trait loci in experiemental crosses*, in *J. Roy. Statist. Soc. B*, 64, pp. 641–656.

Bueso, M. C., Qian, G., and Angulo, J. M. (1999), *Stochastic complexity and model selection from incomplete data*, in *J. Statist. Planning Inference* 76, pp. 273–284.

Cavanaugh, J. E. and Shumway, R. H. (1998), *An Akaike information criterion for model selection in the presence of incomplete data*, in *J. Statist. Planning Inference* 67, pp. 45–65.

Chatterjee, S., Lahiri, P. & Li, H. (2008), *Parametric bootstrap approximation to the distribution of EBLUP, and related prediction intervals in linear mixed models*, in *Ann. Statist.* 36, pp. 1221–1245.

Chen, J. and Chen, Z. (2008), *Extended Bayesian information criteria for model selection with large model spaces*, in *Biometrika* 95, pp. 759–771.

Cisco Systems Inc. (1996), *NetFlow Services and Applications*, White Paper.

Claeskens, G. and Consentino, F. (2008), *Variable selection with incomplete covariate data*, in *Biometrics* 64, pp. 1062–1069.

Craven, P. and Wahba, G. (1979), *Smoothing noisy data with spline functions*, in *Num. Math.* 31, pp. 377–403.

Daiger, S. P., Miller M., and Chakraborty (1984), *Heritability of quantitative variation at the group-specific component (Gc) Locus*, in *Amer. J. Hum. Genet.* 36, pp. 663–676.

Dempster, A., Laird, N., and Rubin, D. (1977), *Maximum likelihood from incomplete data via the EM algorithm (with discussion)*, in *J. Roy. Statist. Soc. B* 39, pp. 1–38.

Datta, G. S. and Lahiri, P. (2001), *Discussions on a paper by Efron & Gous*, in *Model Selection (P. Lahiri ed.), IMS Lecture Notes/Monograph*, Vol. 38, pp. 249–254 (IMS, USA).

Datta, G. S., Hall, P., and Mandal, A. (2011), *Model selection by testing for the presence of small-area effects, and applications to area-level data*, in *J.*

Amer. Statist. Assoc. 106, pp. 361–374.

de Leeuw, J. (1992), *Introduction to Akaike (1973) information theory and an extension of the maximum likelihood principle*, in *Breakthroughs in Statistics* (S. Kotz and N. L. Johnson eds.), Vol. 1, pp. 599–609 (Springer, London).

Diaconis, P. and Mosteller, F. (1989), *Methods for studying coincidences*, in *J. Amer. Statist. Assoc.* 84, pp. 853–861.

Diggle, P. J., Heagerty, P., Liang, K. Y., and Zeger, S. L. (2002), *Analysis of Longitudinal Data* (2nd ed.), Oxford Univ. Press.

Drigalenko, E. (1998), *How sib-pairs reveal linkage*, in *Am. Hum. J. Genet.*, 63, pp. 1242–1245.

Efron, B. (1979), Bootstrap method: Another look at the jackknife, *Ann. Statist.* 7, 1-26.

Efron, B., and Tibshirani, R. J. (1993), *An Introduction to the Bootstrap*, (Chapman & Hall/CRC, New York).

Efron, B., Hastie, T., Johnstone, I. and Tibshirani, R. (2004), *Least angle regression*, in *Ann. Statist.* 32, op. 407–499.

Efron, B. and Tibshirani, R. (2007), On testing the significance of sets of genes, *Ann. Appl. Statist.* 1, 107-129.

Fabrizi, E. and Lahiri, P. (2004), *A new approximation to the Bayes information criterion in finite population sampling*, Tech. Report, (Dept. of Math., Univ. of Maryland)

Fan, J. and Li, R. (2001), *Variable selection via nonconcave penalized likelihood and its oracle properties*, in *J. Amer. Statist. Assoc.* 96, pp. 1348–1360.

Fan, J. and Lv, J. (2008), *Sure independence screening for ultra-high dimensional feature space*, in *J. Roy. Statist. Soc. B* 70, pp. 849–911.

Fan, J. and Lv, J. (2010), *A selective overview of variable selection in high dimensional feature space*, in *Statistica Sinica* 20, pp. 101–148.

Fan, J. and Yao, Q. (2003), *Nonlinear Time Series: Nonparametric and Parametric Methods*, (Springer, New York).

Fay, R. E. and Herriot, R. A. (1979), *Estimates of income for small places: An application of James-Stein procedure to census data*, in *J. Amer. Statist. Assoc.* 74, pp. 269–277.

Friedman, J. (1991), *Multivariate adaptive regression splines (with discussion)*, in *Ann. Statist.*, 19, op. 1–67.

Friedman, J., Hastie, T. and Tibshirani, R. (2008), *Regularization paths for generalised linear models via coordinate descent*, in *J. Statist. Software.* 33, pp. 1–22.

Fu, W. (1998), *Penalized regression: the bridge versus the Lasso*, in *J. Comput. Graph. Statist.* 7, pp. 397–416.

Fuchs, C. (1982), *Maximum likelihood estimation and model selection in contingency tables with missing data*, in *J. Amer. Statist. Assoc.* 77, pp. 270–278.

Fuller, W. A. (2009), *Sampling Statistics*, (Wiley, Hoboken, NJ).

Ganesh, N. (2009), *Simultaneous credible intervals for small area estimation problems*, in *J. Multivariate Anal.* 100, pp. 1610–1621.

Garcia, R. I., Ibrahim, J. G., and Zhu, H. (2010), *Variable selection for regression models with missing data*, in *Statist. Sinica* 20, pp. 149–165.

Ghosh, M., & Rao, J.N.K. (1994), *Small area estimation: An appraisal (with discussion)*, in *Statist. Sci.* 9, pp. 55-93.

Haldane, J. B. S. (1919), *The probable errors of calculated linkage values and the most accurate method of determining gametic from certain zygotic series*, in *J. Genet.* 8, pp. 291–297.

Hand, D. and Crowder, M. (1995), *Practical Longitudinal Data Analysis* , (Chapman & Hall, New York).

Hannan, E. J. (1980), *The estimation of the order of an ARMA process*, in *Ann. Statist.* 8, pp. 1071–1081.

Hannan, E. J. and Quinn, B. G. (1979), *The determination of the order of an autoregression*, in *J. Roy. Statist. Soc. B* 41, pp. 190–195.

Hartley, H. O. and Hocking, R. (1971), *The analysis of incomplete data*, in *Biometrics* 27, pp. 783–823.

Harville, D. A. (1977), *Maximum likelihood approaches to variance components estimation and related problems*, in *J. Amer. Statist. Assoc.* 72, pp. 320–340.

Haseman, J. K., Elston, R. C. (1972), *The investigation of linkage between a quantitative trait and a marker locus*, in *Behavior Genetics*, 2, pp. 3–9.

Hastie, T. J. and Tibshirani, R. J. (1990), *Generalized Additive Models*, (Chapman & Hall, CRC).

Hastie, T., Tibshirani, R., and Friedman, J. (2001), *The Elements of Statistical Learning: Data Mining, Inference, and Prediction*, (Springer, New York).

Hayes, P. M., Liu, B. H., Knapp, S. J., Chen, F., Jones, B., Blake, T., Franckowiak, J., Rasmusson, D., Sorrells, M., Ullrich, S.E., Wesenberg, D., and Kleinhofs, A. (1993), *Quantitative trail locus effects and environmental interaction in a sample of North American barley germ plasm*, in *Theor. Appl. Genet.* 87, pp. 392–401.

Hens, N., Aerts, M., and Molenberghs, G. (2006), *Model selection for incomplete and design-based samples*, in *Statist. Med.* 25, pp. 2502–2520.

Hindorff, L. A., Sethupathy, P., Junkins, H. A., Ramos, E. M., Mehta, J. P., Collins, F. S., and Manolio, T. A. (2009), *Potential etiologic and functional implications of genome-wide association loci for human diseases and traits*, in *Proc. Nat. Acad. Sci.* 106, pp. 9362.

Hu, K., Choi, J., Sim, A., and Jiang, J. (2013), *Best predictive generalized linear mixed model with predictive lasso for high-speed network data analysis*, in *International J. Statist. Probab.*, in press.

Hu, K., Jiang, J., Choi, J., and Sim, A. (2014), *Analyzing high-speed network data*, submitted.

Hunter, D. and Li, R. (2005), *Variable selection using mm algorithms*, in *Ann. Statist.* 33, pp. 1617–1642.

Hurvich, C. M. and Tsai, C.-L. (1989), *Regression and time series model selection in small samples*, in *Biometrika* 76, pp. 297–307.

Ibrahim, J. G., Zhu, H., Carcia, R. I., and Guo, R. (2011), *Fixed and random effects selection in mixed effects models*, in *Biometrics* 67, pp. 495–503.

Ibrahim, J. G., Zhu, H., and Tang, N. (2008), *Model selection criteria for missing-data problems using the EM algorithm*, in *J. Amer. Statist. Assoc.* 103, pp. 1648-1658.

Ishwaran, J. and Rao, J.S. (2005), *Spike and slab variable selection: frequentist and Bayesian strategies*, in *Ann. Statist.*, 33, pp. 730–774.

Jansen, R. C. (1993), *Interval mapping of multiple quantitative trait loci*, in *Genetics* 135, pp. 205–211.

Jansen, R. C., and Stam, P. (1994), *High resolution of quantitative traits into multiple loci via interval mapping*, in *Genetics* 136, pp. 1447–1455.

Jiang, J. (2003), *Empirical method of moments and its applications*, in *Journal of Statistical Planning and Inference* 115, pp. 69–84.

Jiang, J. (2007), *Linear and Generalized Linear Mixed Models and Their Applications*, (Springer, New York).

Jiang, J. (2010), *Large Sample Techniques for Statistics*, (Springer, New York).

Jiang, J. (2005), *Partially observed information and inference about non-Gaussian mixed linear models*, in *Ann. Statist.* 33, pp. 2695–2731.

Jiang, J. (2014), *The fence methods*, in *Advances in Statistics*, Vol. 2014, 1–14, Hindawi Publishing Corp.

Jiang, J. & Lahiri, P. (2006), *Mixed model prediction and small area estimation (with discussion)*, in *TEST* 15, pp. 1–96.

Jiang, J., Lahiri, P. and Wan, S. (2002), *A unified jackknife theory for empirical best prediction with M-estimation*, in *Ann. Statist.* 30, pp. 1782–1810.

Jiang, J., Luan, Y. and Wang, Y. G. (2007), *Iterative estimating equations: Linear convergence and asymptotic properties*, in *Ann. Statist.*, 35, pp. 2233–2260.

Jiang, J., Nguyen, T. and Rao, J. S. (2009), *A simplified adaptive fence procedure*, in *Statist. Probab. Letters* 79, pp. 625–629.

Jiang, J., Nguyen, T. and Rao, J. S. (2010), *Fence method for nonparametric small area estimation*, in *Survey Methodology* 36, pp. 3-11.

Jiang, J., Nguyen, T. and Rao, J. S. (2011a), *Best predictive small area area estimation*, in *J. Amer. Statist. Assoc.* 106, pp. 732–745.

Jiang, J., Nguyen, T. and Rao, J. S. (2011), *Invisible fence methods and the identification of differentially expressed gene sets*, in *Statist. Interface* 4, pp. 403–415.

Jiang, J., Nguyen, T. and Rao, J. S. (2015), *The E-MS algorithm: Model selection with incomplete data*, in *J. Amer. Statist. Assoc.*, in press.

Jiang, J. and Rao, J.S. (2003), *Consistent procedures for mixed linear model selection*, in *Sankhya A* 65, pp. 23–42.

Jiang, J., Rao, J. S., Gu, Z. and Nguyen, T. (2008), *Fence methods for mixed model selection*, in *Ann. Statist.* 36, pp. 1669–1692.

Jiang, J. and Zhang, W. (2001), *Robust estimation in generalized linear mixed models*, in *Biometrika* 88, pp. 753–765.

Jones, R.H., (1993) *Longitudinal Data with Serial Correlation A State-Space Approach*, (Chapman & Hall, New York).

Karim, M. R. and Zeger, S. L. (1992), *Generalized linear models with random effects: Salamander mating revisited*, in *Biometrics* 48, pp. 631–644.

Kauermann, G. (2005), *A note on smoothing parameter selection for penalized spline smoothing*, in *J. Statist. Planning & Inference* 127, pp. 53–69.

Lander, E. S. and Botstein, D. (1989), *Mapping mendelian factors underlying quantitative traits using RFLP linkage maps*, in *Genetics* 121, pp. 185–199.

Lange, K. (2002), *Mathematical and Statistical Methods for Genetic Analysis*, (2nd ed.), (Springer, New York).

Lehmann, E. L. (1986), *Testing Statistical Hypotheses*, (2nd ed.), (Chapman & Hall, London).

Lehmann, E. L. (1999), *Elements of Large-Sample Theory*, (Springer, New York).

Liang, K. Y. and Zeger, S. L. (1986), *Longitudinal data analysis using generalized linear models*, in *Biometrika* 73, pp. 13–22.

Lin, X. and Breslow, N. E. (1996), *Bias correction in generalized linear mixed models with multiple components of dispersion*, in *J. Amer. Statist. Assoc.* 91, pp. 1007–1016.

Little, R. J. A. and Rubin, D. B. (2002), *Statistical Analysis with Missing Data*, 2nd ed., (Wiley, New York).

Monte Carlo Strategies in Scientific Computing, (Springer, New York).

Liu, R. Y. (1988), *Bootstrap procedures under some non-i.i.d. models*, in *Ann. Statist.* 16, pp. 1696–1708.

Luenberger, D. G. (1984), *Linear and Nonlinear Programming*, (Addison-Wesley, Reading, MA).

Luo, Z. W. *et al.* (2007), *SFP genotyping from affymetrix arrays is robust but largely detects cis-acting expression regulators*, in *Genetics* 176, pp. 789–800.

McCullagh, P. and Nelder, J. A. (1989), *Generalized Linear Models*, (2nd ed.), (Chapman & Hall, London).

Meier, L., van de Geer, S., and Bühlmann, P. (2008), *The group lasso for logistic regression*, in *J. Roy. Statist. Soc.* 70, pp. 53–71.

Melcon, E. (2014), *On optimized shrinkage variable selection in generalized linear models*, *J. Statist. Comput. Simulation*, revised.

Meza, J. & Lahiri, P. (2005), *A note on the Cp statistic under the nested error regression model*, in *Survey Methodology* 31, pp. 105–109.

Miller, J. J. (1977), *Asymptotic properties of maximum likelihood estimates in the mixed model of analysis of variance*, in *Ann. Statist.* 5, pp. 746–762.

Morris, C. N. & Christiansen, C. L. (1995), *Hierarchical models for ranking and for identifying extremes with applications*, in *Bayes Statistics* 5, (Oxford Univ. Press).

Mou, J. (2012), Two-stage fence methods in selecting covariates and covariance for longitudinal data, Ph. D. dissertation, Dept. of Statist., Univ. of Calif., Davis, CA.

Müller, S., Scealy, J. L., and Welsh, A. H. (2013), *Model selection in linear mixed models*, in *Statist. Sci.* 28, pp. 135–167.

Münnich, R., Burgard, J. P., and Vogt, M. (2009), *Small area estimation for population counts in the German Census 2011*, in *Section on Survey Research Methods*, JSM 2009, (Washington, D.C.)

Nakamichi, R., Ukai, Y., and Kishino, H. (2001), *Detection of closely linked multiple quantitative trait loci using genetic algorithm*, in *Genetics* 158, pp. 463–475.

Nguyen, T. and Jiang, J. (2012), *Restricted fence method for covariate selection in longitudinal data analysis*, in *Biostatist.* 13, pp. 303–314.

Nguyen, T., Peng, J., and Jiang, J. (2013), *Fence methods for backcross experiments*, in *J. Statist. Comput. Simulations*, in press.

Nishii, R. (1984), *Asymptotic properties of criteria for selection of variables in multiple regression*, in *Ann. Statist.* 12, pp. 758–765.

Opsomer, J. D., Breidt, F. J., Claeskens, G., Kauermann, G. and Ranalli, M. G. (2008), *Nonparametric small area estimation using penalized spline regression*, in *J. Roy. Statist. Soc. B* 70, pp. 265–286.

Osborne, M. R., Presnell, B. and Turlach, B. A. (2000), *On the LASSO and its dual*, in *J. Comput. Graph. Statist.* 9, pp. 319–337.

Owen, A. (2007), *The pigeonhole bootstrap*, Tech. Report, Dept of Stat., Stanford Univ.

Pang, Z., Lin, B. and Jiang, J. (2013), *Regularization parameter selections with divergent and NP-dimensionality via bootstrapping*, in *Austral. New Zealand J. Statist.*, in press.

Park, M. Y. and Hastie, T. (2007). *L_1-regularization path algorithm for generalized liner models*, in *J. Roy. Statist. Soc. B* 69, pp. 659-677.

Pebley, A. R., Goldman, N. and Rodriguez, G. (1996), *Prenatal and delivery care and childhood immunization in Guatamala; do family and community matter?*, in *Demography* 33, pp. 231–247.

Piepho, H.-P. and Gauch, H. G. (2001), *Marker pair selection for mapping quantitative trait loci*, in *Genetics* 157, pp. 433–444.

Rao, J. S. (1999), *Bootstrap choice of cost complexity for better subset selection*, in *Statistica Sinica* 9, pp. 273–288.

Rao, J. N. K. (2003), *Small Area Estimation*, (Wiley, Hoboken, NJ).

Rissanen, J. (1983), *A universal prior for integers and estimation by minimum description length*, in *Ann. Statist.* 11, pp. 416–431.

Rodriguez, G. and Goldman, N. (2001), *Improved estimation procedure for multilevel models with binary responses: A case-study*, in *J. Roy. Statist. Soc. A* 164, pp. 339–355.

Robins, J. M., Rotnitzky, A. and Zhao, L. P. (1995), *Analysis of semiparametric regression models for repeated outcomes in the presence of missing data*, in *J. Amer. Statist. Assoc.* 90, pp. 106–121.

Robinson, G. K. (1991), *That BLUP is a good thing: The estimation of random effects (with discussion)*, in *Statist. Sci.* 6, pp. 15–51.

Rotnitzky, A., Robins, J. M. and Scharfstein, D. (1998), *Semiparametric regression for repeated outcomes with nonignorable nonresponses*, in *J. Amer. Statist. Assoc.* 93, pp. 1321–1339.

Rubin, D. B. (1976), *Inference and missing data*, in *Biometrika* 63, pp. 581–592.

Ruppert, R., Wand, M. and Carroll, R. (2003), *Semiparametric Regression*, (Cambridge Univ. Press).

Sebastiani, P. and Ramoni, M. (2001), *Bayesian selection of decomposable models with incomplete data*, in *J. Amer. Statist. Assoc.* 96, pp. 1375–1386.

Sen, A. and Srivastava, M. (1990), *Regression Analysis*, (Springer, New York).

Schomaker, M., Wan, A. T. K., and Heumann, C. (2010), *Frequentist model averaging with missing observations*, in *Comput. Statist. Data Anal.* 54, pp. 3336–3347.

Schwarz, G. (1978), *Estimating the dimension of a model*, in *Ann. Statist.* 6, pp. 461–464.

Seghouane, A.-K., Bekara, M., and Fleury, G. (2005), *A criterion for model selection in the presence of incomplete data based on Kullback's symmetric divergence*, in *Signal Process.* 85, pp. 1405–1417.

Shao, J. (1993), *Linear model selection by cross-validation*, in *J. Amer. Statist. Assoc.* 88, pp. 486–494.

Shao, J. (1996), *Bootstrap model selection*, in *J. Amer. Statist. Assoc.* 91, pp. 655–665.

Shao, J. and Rao, J.S. (2000), *The GIC for model selection: a hypothesis testing approach*, in *J. Statist. Plann. Inference* 88, pp. 215–231.

Shibata, R. (1976), *Selection of the order of an autoregressive model by Akaike's information criterion* in *Biometrika* 63, pp. 117–126.

Shibata, R. (1984), *Approximate efficiency of a selection procedure for the number of regression variables*, in *Biometrika* 71, pp. 43–49.

Shimodaira, H. (1994), *A new criterion for selecting models from partially observed data*, in P. Cheeseman and R. W. Oldford, eds., *Selecting Model from Data: Artificial Intelligence and Statistics IV, Lecture Notes in Statistics* 89, (Spriner, New York), 21–29.

Shorack, G. R. and Wellner, J. A. (1986), *Empirical Processes with Applications to Statistics*, (Wiley, New York).

Shumway, R. H. (1988), *Applied Statistical Time Series Analysis*, (Prentice-Hall, Englewood Cliffs, NJ).

Siegmund, D. and Yakir, B. (2007) *The Statistics of Gene Mapping*, (Springer, New York).

Silanpää, M. J. and Corander, J. (2002), *Model choice in gene mapping: what and why*, in *Trends in Genetics* 18, pp. 301–307.

Subramanian, A., Tamayo, P., Mootha, V. K., Mukherjee, S., Ebert, B. L., Gillette, M. A., Paulovich, A., Pomeroy, S. L., Golub, T. P., Lander, E. S. and Mesirov, J. P. (2005), Gene set enrichment analysis: A knowledge-based approach for interpreting genome-wide expression profiles, *Proc. Natl. Acad. Sci. USA* 102, 15545-15550.

Tibshirani, R. J. (1996), *Regression shrinkage and selection via the Lasso*, in *J. Roy. Statist. Soc. Ser. B* 16, pp. 385–395.

Tibshirani, R. J. (2005), *A simple explanation of the lasso and least angle regression*, www-stat.stanford.edu/~tibs/lasso/simple.html.

Torabi, M. (2012), *Likelihood inference in generalized linear mixed models with two components of dispersion using data cloning*, in *Comput. Statist. Data Anal.* 56, pp. 4259–4265.

van Buuren, S., Brand, J. P. L., Groothuis-Oudshoorn, C. G. M., and Rubin, D. B. (2005), *Fully conditional specification in multivariate imputation*, in *J. Statist. Comput. Simulation* 76, pp. 1049–1064.

Verbeke, G., Molenberghs, G., and Beunckens, C. (2008), *Formal and informal model selection with incomplete data*, in *Statist. Sci.* 23, pp. 201–218.

Wang, H., Li, R. and Tsai, C. L. (2007), *On the consistency of SCAD tuning parameter selector*, in *Biometrika* 94, pp. 553–558.

Wright, F. (1997), *The phenotypic difference discards sub-pair QTL linkage information*, in *Am. J. Hum. Genet.*, 60, pp. 740–742.

Wu, C. F. J. (1986), *Jackknife, bootstrap and other resampling methods in regression analysis*, in *Ann. Statist.* 14, pp. 1261–1295.

Ye, J. (1998), *On measuring and correcting the effects of data mining and model selection*, in *J. Amer. Statist. Assoc.*, 93, pp. 120–131.

Zhan, H., Chen, X., and Xu, S. (2011), *A stochastic expectation and maximization algorithm for detecting quantitative trait-associated genes*, in *Bioinformatics* 27, pp. 63–69.

Zhang, Y., Li, R. and Tsai, C. L. (2010), *Regularization parameter selections via generalised information criterion*, in *J. Amer. Statist. Assoc.* 105, pp. 312–323.

Zeng, Z-B. (1993), *Theoretical basis of separation of multiple linked gene effects on mapping quantitative trait loci*, in *Proc. Natl. Acad. Sci.* 90, pp. 10972–10976.

Zeng, Z-B. (1994), *Precision mapping of quantitative trait loci*, in *Genetics* 136, pp. 1457–1468.

Zou, H. (2006), *The adaptive Lasso and its oracle properties*, in *J. Amer. Statist. Assoc.* 101, pp. 1418–1429.

Zou, H., and Hastie, T. (2005). *Regularization and Variable Selection via the Elastic Net*, *J. Roy. Statist. Soc. Ser. B* 67 , pp. 301–320.

Zou, H. and Li, R. (2008), *One-step sparse estimates in nonconcave penalized likelihood models*, in *Ann. Statist.* 36, pp. 1509-1533.

Zou, H. and Zhang, H. (2009), *On the adaptive elastic-net with a diverging number of parameters*, in *Ann. Statist.* 37, pp. 1733–1751.

Index

www.ingramcontent.com/pod-product-compliance
Lightning Source LLC
Chambersburg PA
CBHW050555190326
41458CB00007B/2052